Animals, Anthropomorphism and Mediated Encounters

This book critically investigates the pervasiveness of anthropomorphised animals in popular culture.

Anthropomorphism in popular visual media has long been denounced for being unsophisticated or emotionally manipulative. It is often criticised for over-expressing similarities between humans and other animals. This book focuses on everyday encounters with visual representations of anthropomorphised animals and considers how attributing other animals with humanlike qualities speaks to a complex set of power relations. Through a series of case studies, it explores how anthropomorphism is produced and circulated and proposes that it can serve to create both misunderstandings and empathetic connections between humans and other animals.

This book will appeal to academics and students interested in visual media, animal studies, sociology and cultural studies.

Claire Parkinson is Professor of Film, Television and Digital Media and Co-director of the Centre for Human Animal Studies at Edge Hill University. Her research interests cover media, film and animal studies. Her publications include the books *Popular Media and Animals* and *Beyond Human*.

Routledge Human–Animal Studies Series

Series edited by Henry Buller

Professor of Geography, University of Exeter, UK

The new *Routledge Human–Animal Studies Series* offers a much-needed forum for original, innovative and cutting-edge research and analysis to explore human–animal relations across the social sciences and humanities. Titles within the series are empirically and/or theoretically informed and explore a range of dynamic, captivating and highly relevant topics, drawing across the humanities and social sciences in an avowedly interdisciplinary perspective. This series will encourage new theoretical perspectives and highlight ground-breaking research that reflects the dynamism and vibrancy of current animal studies. The series is aimed at upper-level undergraduates, researchers and research students as well as academics and policy-makers across a wide range of social science and humanities disciplines.

Animal Housing and Human–Animal Relations
Politics, Practices and Infrastructures
Edited by Kristian Bjørkdahl and Tone Druglitrø

Shared Lives of Humans and Animals
Animal Agency in the Global North
Edited by Tuomas Räsänen and Taina Syrjämaa

(Un)Stable Relations
Horses, Humans and Social Agency
Lynda Birke and Kirrilly Thompson

Carceral Space, Prisoners and Animals
Karen M. Morin

Historical Animal Geographies
Edited by Sharon Wilcox and Stephanie Rutherford

Animals, Anthropomorphism and Mediated Encounters
Claire Parkinson

For a full list of titles in this series, please visit: www.routledge.com/Routledge-Human-Animal-Studies-Series/book-series/RASS

Animals, Anthropomorphism and Mediated Encounters

Claire Parkinson

Routledge
Taylor & Francis Group

LONDON AND NEW YORK

First published 2020
by Routledge
2 Park Square, Milton Park, Abingdon, Oxon OX14 4RN

and by Routledge
605 Third Avenue, New York, NY 10017

First issued in paperback 2021

Routledge is an imprint of the Taylor & Francis Group, an informa business

British Library Cataloguing-in-Publication Data
A catalogue record for this book is available from the British Library

Library of Congress Cataloging-in-Publication Data
Names: Parkinson, Claire, author.
Title: Animals, anthropomorphism and mediated encounters/
    Claire Parkinson.
Description: Milton Park, Abingdon, Oxon ; New York, NY : Routledge,
    2020. | Series: Routledge human-animal studies |
    Includes bibliographical references and index.
Identifiers: LCCN 2019015165 | ISBN 9780367195731 (hbk) |
    ISBN 9780429203244 (ebk)
Subjects: LCSH: Animal psychology. | Anthropomorphism. |
    Human-animal relationships.
Classification: LCC QL785. P24 2020 | DDC 591.5—dc23
LC record available at https://lccn.loc.gov/2019015165

ISBN 13: 978-0-367-78529-1 (pbk)
ISBN 13: 978-0-367-19573-1 (hbk)

Typeset in Times New Roman
by Apex CoVantage, LLC

For Trinny and Logan

Contents

Acknowledgements

This book was written with financial support from the Edge Hill University Research Investment Fund.

I am grateful to the many academics who read the proposal for this book as well as various drafts of individual chapters and the draft manuscript. In some cases, these have been anonymous reviewers and I want to thank them all for their insights and constructive comments. Some of the ideas here have been road tested as papers and talks and I'm grateful to all those who have listened to me, asked challenging questions and given thoughtful feedback. Thanks to my wonderful family, human and nonhuman, for their love and unwavering support.

1   Introduction

Anthropomorphism is ubiquitous in western popular culture. In general, anthropomorphism refers to the attribution of human characteristics or behaviour to a god, 'animal' or object, and it is the second of these categories – 'animal' – which is arguably the most contentious. Such is its prevalence that anthropomorphism shapes ideas about nonhuman animals more than any other aspect of their popular representation. Yet, despite their undeniable and enduring popularity, anthropomorphised animals are considered a problem. The overexpression of similitude between humans and other animals has become synonymous with Disney and a set of representational practices apparent in popular culture that reduce other species to simple feathered, furred and scaled human analogues. Anthropomorphised animals are, so the argument might go, subsumed into a human social logic where their commodification, especially for a family audience, is predicated on the erasure of their individual complexity and species difference. In its pejorative sense, anthropomorphism remains to some extent weighed down with associations to childishness, a lack of objectivity and sentimentality.

Concerns about anthropocentric conceit motivate valid criticisms of anthropomorphism, when it serves only, or primarily, human interests. Kari Weil observes that 'as a process of identification, the urge to anthropomorphize the experience of another, like the urge to empathize with that experience, risks becoming a form of narcissistic projection that erases boundaries of difference' (Weil, 2012: 19). Weil draws attention to the importance of difference in critical discussions about anthropomorphism. The stakes are high, and in humanising animals we risk losing sight of them as beings in their own right, with individual experiences and capacities that are quite different from our own. But there are also good reasons to be critical of the rejection of anthropomorphism where it is also motivated by anthropocentric concerns that sustain oppression, exploitation and suffering. As Richard Ryder remarks, 'The words "anthropomorphism" and "sentimentality", both widely used in twentieth century Britain to disparage those who treated nonhuman animals in ways considered to be only appropriate to humans, were unheard in this context until after Darwin's day'. He goes on to ask, 'Is it too fanciful to suggest that they were the animal exploiter's defences against the logical implications of Darwinism?' (Ryder, 1989: 164). Why should it be that the use of animals as human proxies in biomedical research is not considered anthropomorphic, yet challenges to such practices on the basis that those individual beings are sentient and suffer can be dismissed as sentimental anthropomorphism? Ryder's question throws anthropomorphism into the centre of the politics of human-animal relations. It also raises a further question about what anthropomorphism has the potential to do *for* nonhuman animals. In this sense, anthropomorphism is a disruptive force, a capacity for imaginative appreciation of another's perspective; it opens the opportunity for cross-species intersubjectivity, and it can play a role in the development of empathetic relationships with other animals. Studies which suggest that anthropomorphism is linked to pro-environmental behaviours (Tam, 2015; Waytz et al., 2010) and that a tendency to anthropomorphise is related to lowered meat consumption and increased concern for the welfare of animals (Niemyjska et al., 2018) would seem to support the contention that anthropomorphism has a role to play in expanding effective forms of human concern for the wellbeing of other animals. At the core of this book is the claim that anthropomorphised animals in popular culture are highly significant within the politics of human-animal relations and intervene in discourses that shape the practices which govern the material lives of species other than humans. Crucially, anthropomorphism in popular culture engages both human empathy for and misunderstanding of other animals.

Mediation and media

The term anthropomorphism is derived from the Greek *anthrōpos* (human) and *morphē* (form). Until the latter half of the nineteenth century, it referred to the practices of attributing deities with humanlike characteristics or bodily form. By the first decades of the twentieth century, anthropomorphism had come to be regarded, in a pejorative sense, as the attribution of uniquely human characteristics to other animals. The widespread rejection of anthropomorphism within western science and 'serious' art exposed various fracture points in humanism that assembled around the threat to the unified rational subject conceived of in terms of human exceptionalism. Anthropomorphism, in other words, can be a troubling irritant to ideas of human uniqueness. Allied with this, the feminisation of emotion and sentiment, the hierarchisation of knowledges that dismissed forms of animism as primitive and anti-enlightenment, and ambivalence towards popular culture also played a role in the legitimacy afforded to *non*-anthropomorphic ways of thinking about, seeing and representing animals and their experiences. If we scratch at the surface of the historical regulation of anthropomorphism, we find that it has been closely managed for more than a century by anthropocentric ideas and racialised and gendered systems of thought. Despite such endeavours, anthropomorphism has not been expunged from our contemporary lives. On the contrary, anthropomorphism and its equally problematic (for some) accomplice, sentimentalism, are alive and well and circulating in abundance throughout systems of cultural production.

This book is primarily concerned with anthropomorphism as it relates to mediated encounters with other animals. With the focus on popular culture, I use the term 'mediated encounter' to signal not only the emotional appeals and audience reception of nonhuman animal representations but also the processes and practices by which individual sentient beings are produced as anthropomorphic, commodified narrative agents. This means that, at times, I give space to discussions of the industrial and cultural practices and conventions involved in creating film, television, social media content, advertising and so forth and the discourses that shape the mediated encounter. I do this because it is vital that we acknowledge not only the potential for stories to mobilise empathy and misunderstandings about other animals, but to also recognise that those narratives are constructed through practices and processes that involve real animals. This in turn raises ethical questions about their treatment within systems of cultural production and the asymmetries of power that are involved.

By thinking about the mediated encounter, I move away from a concern with representations of other animals as having only or mainly symbolic value in service to our understanding of human identity. Instead, I take the mediated encounter as the meeting point between the institutional, social and industrial practices and processes that reshape nonhuman animals into commodified narrative agents, the affective dimensions and emotional appeals that are involved and the reception of such encounters by human audiences. In the human-to-human communication of popular culture, mediation is an act that brings 'the animal' into a human world. Mediation in this sense is inevitably anthropomorphic, and it is the constellation of conditions and relations that relate to it which concern the substantive focus of this book.

The mediated encounter does not exclude embodied encounters. Filming, photographing or otherwise recording another animal involves a material encounter between humans, technologies and a nonhuman animal subject, and across the book I discuss anthropomorphism in relation to this wider understanding of mediation as process. At the same time, it is crucial to point out that the direct embodied and mediated encounter differ and to acknowledge that while they are relational (in that the material encounter is reshaped by and as the mediated encounter), they are not interchangeable. Cultural production and consumption encompass myriad situated practices that are entangled but, at the risk of overstating the obvious, the embodied 'real-life' encounter with another animal is qualitatively and materially different to a mediated encounter via a screen with that same individual animal. This should not be taken to imply that we arrive at the embodied encounter without some form of mediating knowledge or setting; the spaces in which we encounter animals are always highly organised and their meanings managed. I start from the position that the processes of mediation are always situated, privilege certain senses, viewpoints, knowledges and ideologies and rely on specific institutionalised practices, strategies of engagement and systems of meaning. Nor does this approach imply that we are not embodied sensorial subjects when we engage with onscreen media. On the contrary, as Vivien Sobchack contends, when we 'watch' onscreen media, we do not experience it only through our eyes, we 'see and comprehend and feel [. . .] with our entire bodily being, informed by the full history and carnal knowledge of our acculturated sensorium' (Sobchack, 2004:

63). This then prompts the question, how does mediation specific to the cultural production and circulation of anthropomorphic animal stories and images shape our encounters with other species?

To answer this question, my approach is informed by the shift in media studies towards understanding cultural production and consumption as situated practices, a form of 'radical contextualisation' (Ang, 1996) that moves beyond 'the text' – the individual film, programme, image and so forth – to a thick context that encompasses practices of production, aspects of reception and affect, the text and its entanglements with other narratives, technologies and cultural forms (Spitulnik, 2010). Within media studies, such an approach commonly focuses on media entanglements in the lives of people where media is seen primarily in terms of human production, distribution, circulation and consumption (Spitulnik, 2010: 105; Merskin, 2016: 16). As such there has been a blind spot when it comes to animals even though species other than humans are central to so much popular media content. In recent years, animal studies and critical animal studies (CAS) have found disciplinary alliances within film, media, communications and cultural studies such that there has been a greater scholarly focus on nonhuman animals and the development of a discrete sub-field: 'critical animal and media studies' (Almiron and Cole, 2016). It is nonetheless the case that 'animals' remain a specialist field of interest within the larger disciplines and where they do figure as central in scholarly work, differences in approach reflect varying levels of concern with their treatment and the role of media in sustaining normative views and attitudes about animal issues.

Almiron and Cole contrast animal studies with critical animal studies approaches in media and communications scholarship, arguing that the former holds a more mainstream position that although acknowledging the central importance of nonhuman animals does not challenge their oppression, while the latter position aligns strongly with an anti-speciesist praxis (Almiron and Cole, 2016: 3). Crucially, the obligations of critical animal and media studies push to the foreground questions of the relation between cultural representations and the material conditions of actual animals (Molloy, 2011; Merskin, 2016: 19). Such an impetus can drive academic enquiry to look at connected systems and practices and the messy configurations that develop via social, industrial, cultural and economic discursive formations (Molloy, 2011). To these ends, in its aims to address issues of oppression, domination, control and power, critical animal studies finds an intellectual ally in the political economy of communication and the wider concerns of Critical Theory, cultural studies and critical media studies. For this reason, the convergence between critical animal studies and critical media studies, Almiron and Cole suggest, was inevitable (Almiron and Cole, 2016: 2).

Informed by critical animal studies, I look to popular media forms to ask if and how anthropomorphism can be utilised *effectively* to mobilise empathy for other animals. As Almiron and Cole point out, to claim a critical animal studies position is to demarcate a specific set of intentions and motivations. The anti-speciesist praxis that they propose is usefully summarised by Nik Taylor and Richard Twine (2015: 2) as 'concerned with the nexus of activism, academia and animal suffering

and maltreatment. [. . .] CAS takes a normative stance against animal exploitation and so "critical" also denotes a stance against an anthropocentric status-quo
in human-animal relations'. A CAS approach then shapes the intent to untangle
anthropomorphism as a form of anthropocentric projection *and* its rejection where
it is underpinned by anthropocentric interests that sustain animal exploitation from
anthropomorphism as an affective and effective means of mobilising empathy. In
doing so I pay attention to ways in which gendered norms come to inform our
understanding of other animals and where they are deployed to deny, reduce or erase
the experiences of real animals. In this regard, I acknowledge the extent to which
pejorative meanings ascribed to anthropomorphism have relied on dualisms such
as the feminisation of emotion opposed with a masculinised discourse of scientific
objectivity which has distanced itself from associations with 'primitive' animism.

Despite the problematisation of 'objective' scientific knowledge by feminist
scholars (Haraway, 1988, 1989, 1991), discourses of science continue to haunt
our understanding and critiques of anthropomorphism. Much of the twentieth
century was characterised by the authority of scientific knowledge production,
capitalism and humanism within a mutually informing discursive formation that
has been, in large part, predicated upon differentiating humanness from animality
to privilege human progress. A consequence of maintaining difference between
humans and other animals has been to deny the validity of anthropomorphic practices and thereby render it a casualty of the politics of human-animal difference.
Shifts in thinking about animals, similitude and difference occur across the twentieth and twenty-first centuries and as a result discourses correspondingly flex
and alter according to knowledge conditions, revising and redacting the story of
human-animal relations and anthropomorphism. These revisions change the ways
in which we can meaningfully represent other species and talk about our relations
with them. In this regard, then, where I turn my attention to instances of anthropomorphism, I am interested in reading them not simply as media representations
of anthropomorphic animals but as social and cultural practices of making other
animals 'like us', with all the complications that that entails.

The expanded context approach taken here understands anthropomorphism as
situated and culturally contingent, but in adopting a 'wide-angle lens on media'
(Spitulnik, 2010) the thorny question of how the object of study is delimited
becomes all the more urgent. Expanded approaches can move in multiple directions, and the ubiquity of mediated encounters with anthropomorphised animals
presents a further problem in terms of narrowing the field, and within that lurks
another issue, that of the homogenising category of 'animal'.[1] To address these
issues, I take a thematic approach to anthropomorphism generally with Chapters 3 through 6, being concerned with different ideological boundaries between
humans and other animals that have informed western critiques of anthropomorphic practice. Chapter 3 deals with seeing and the relationship between human
sensory hierarchies and ethical concerns; Chapter 4 explores emotion; Chapter 5
looks at language; and Chapter 6 focuses on the thinking mind. The cultural con-
tingency of these ideological boundaries and the situated practices that they frame
as anthropomorphic delimit the scope of this study to western media.[2]

'Animal', as a category, homogenises a massive diversity of life and beings, and in attending to anthropomorphism as a situated practice it has been necessary to be selective in the number and range of case studies. Chapter 3 explores the mediation of animal experience through a study of praying mantises, conceived here as a limit case to test out the arbitrary ethical lines that humans draw in relation to our concern for other species. Chapter 4 explores the ways in which emotion is ascribed or not in particular instances of photography, social media, news and advertising to free-roaming (wild), companion (pet) and farmed animals through case studies of kangaroos, dogs and cows. Chapter 5 turns to the topic of language in relation to aquatic mammals and dogs, looking at the cultural sites where science and fiction collide. Chapter 6 then examines nonhuman animal minds and how film and television communicate the capacities of sharks and orca for mental suffering. Anthropomorphic practices are diverse and not all animals are anthropomorphised in the same way. The aims here are therefore to move beyond thinking about a singular monolithic anthropomorphism and the arguments around category error or methodological validity, to acknowledge anthropomorphism's entangled relations as a situated practice, and to explore its potential to contribute to the development of empathetic human relationships with other animals.

An affective/discursive approach

In one sense, there is a problem inherent in the aims I mention above, and that is that the discourses that continue to treat anthropomorphism as a methodological issue or category error also continue to shape, usually by some act of circumscription, the situated practices of anthropomorphism (Molloy, 2001). To examine anthropomorphism as a situated practice by an expanded approach necessarily involves acknowledging the discourses that shape its meaning in context. Where anthropomorphism is contested as a practice, we find sites where boundary maintenance between humans and other animals is negotiated within popular culture. These are spaces where the discontinuity discourse that views humans and other animals as absolutely distinct intervenes to 'correct' popular anthropomorphic interpretation (such as I discuss in Chapter 4). Often this will be in the form of media accounts of some public 'error' of anthropomorphic interpretation that 'needs' to be corrected by an objective scientific explanation. In this way, knowledge about other animals is mediated by culture and 'scientific objectivity' is reasserted as 'common sense'. Nik Taylor draws attention to the problems of scientific circumscription and suggests that, to get past the debates about whether anthropomorphism is valid or not, we must recognise and deconstruct the discourses of science that declare anthropomorphism to be illegitimate *and* look at the actual practices of anthropomorphism (Taylor, 2011: 268). Richie Nimmo similarly questions why it should be that, when it comes to anthropomorphism, we normalise scientific scepticism as a measure of how we should judge our interactions with other animals (Nimmo, 2016: 18). Such questions call attention to discourses as historically situated knowledge and bring to the foreground the ongoing contest between different knowledge forms that frame our ideas of discontinuity or continuity between humans and other animals.

One approach to understanding the relationship between knowledge and power argues that the identification of discontinuity defines 'the mode of being of the objects that appear in that field' and sustains the conditions by which a discourse about them is recognised as truthful (Foucault, 1997: 158). In general, this suggests that the organisation of difference is central to sustaining particular truths within a society and validating power relations. Difference is not, however, only about constructing otherness and while it can be a source of disunity and fragmentation; difference can also mobilise resistance and change. It is not my intention here to discourage attention to species difference or indeed to argue for human similitude to other species as the grounds for their better treatment. In this book I am, however, interested in how the understanding of difference and similitude is deployed in human relationships with other animals and the ways in which that impacts on their lives and experiences.

How anthropomorphism is discursively shaped, the places where discourses overlap and reinforce each other, and the places where they compete and open new sites for hegemonic practices to be contested are important to this discussion. Across this book, I turn my attention to how anthropomorphism is defined and deployed in different conceptual and physical spaces, how it draws on other discourses of animality, difference, similitude, gender and so forth to reinforce 'truths' about animals, where prevailing norms are contested and what that means for the material lives of other animals. To do this, I draw on an affective/discursive approach which takes a broad view of discourse as being not only linguistic but multimodal, relating to sound, visuality, writing and so forth. It is important to distinguish this use of the term discourse here as a narrower definition refers specifically to spoken or written language (see, for example, Fairclough, 1992). Here discourses are not taken to be confined to language statements but are instead active across the various modalities of popular culture. I am therefore attentive to the processes of delimiting, designating, naming and establishing anthropomorphism as an object of discourse, the sites where popular culture figures as a mediator of the knowledge that shapes anthropomorphic practices, how claims to truth in the interpretation of our encounters with nonhuman animals are contested and how these various intersecting relations impact on the material lives of real nonhuman animals. These processes relate to how power operates, how anthropomorphism becomes invested with specific meanings and where these translate into 'truths' about animals. In other words, who decides where, how and when anthropomorphism is valid or not and what types of animal representations, practices and treatment are endorsed as a consequence?

Affect and empathy

In terms of affective engagement, I am interested in how affect is mediated and culturally produced and how discourses shape affective responses. In keeping with other general definitions, affect is taken to refer here to be an embodied visceral response while emotion is considered the cognitive awareness of affect (Nyman and Schuurman, 2016: 2). I acknowledge that affects are intertwined with discourses in complex ways, for instance, they might 'reproduce or strengthen discourses, sometimes they open possibilities for future changes, and sometimes they motivate discursive crisis' (Knudsen and Stage, 2015: 20). Bodies affect and are affected, and this focus has proved to be an especially important 'turn' for those interested in human-animal relations not least because affect is 'in excess of the practices of the 'speaking subject'' (Blackman and Venn, 2010: 9). Undoing the centrality of the rational, speaking human subject that in traditional binaries is opposed to the animal object, the affective turn enables an acknowledgement that humans and other animals are affected by one another, a relationality that posits a different type of connection between interacting entities. In the mediated encounter, anthropomorphism can reproduce an animal's body as a site of affect, an opportunity for cross-species intersubjectivity and an imaginative shift to a shared embodied experience. The imaginative shift that I refer to here reflects Matthew Calarco's proposition that 'poetic, literary, or artistic descriptions of animals [. . .] might help us to see animals otherwise, which is to say, otherwise than the perspectives offered by the biological sciences, common sense, or the anthropocentric "wisdom" of the ages' (Calarco, 2008: 127). However, the relationality of such an imaginative shift must, I contend, be overlaid with the economic realities of cultural production. The affective relations of anthropomorphism draw attention to the affective labour that other animals do, and it is crucial that we recognise how that impacts on the lived material lives of real nonhuman animals. In this book I explore how anthropomorphism figures in the entanglement of neoliberal economics and affective management and where it – as is the case with other forms of affective labour – acts as a salve for human anxieties under the rubric of neoliberal precarity. It can also mobilise empathetic connections and, if we take seriously the entanglements of anthropomorphism, the two, I argue, are not mutually exclusive. Moreover, I propose that where the corporate production of culture relies on the economic exploitation of anthropomorphic animals, this exposes a potential vulnerability to reputational capital, a matter which I discuss in relation to case studies of the 'Blackfish effect' and the promotion of farmed animal products. In this sense, I propose that anthropomorphism is used to distance human consumers from the realities of animal cruelty and exploitation, but that distancing strategy relies on reinforcing sympathetic connections between human and nonhuman animal. In this book I offer a critique of selected commercial strategies that encourage anthropomorphic identifications, deployed to construct an illusory sense of nonhuman animal autonomy and agency.

As I am interested in the sites where anthropomorphism mobilises empathy, it is important to acknowledge that a discourse, critical of precisely this relationship between anthropomorphism and emotion, already circulates. Indeed, despite the affective turn in academia, a prevailing criticism of anthropomorphism is that it is 'merely' a form of sentimentality – a self-indulgent expression of misplaced emotion towards other species. In literature, the tradition of sentimentalism has been thought of as a 'lesser' form produced by women or popular writers that, through convention, stock characters and rhetorical devices aimed to arouse the emotions (Nyman, 2016: 66). In cinema, the alignment of sentimentalism with women's films relegated their emotional appeals to 'low' or second-rate cultural forms that were considered 'in excess' of the boundaries of serious or respectable drama. Contemporary thinking about sentimentalism still maintains a belief that it is 'an outdated mode' that 'portrays emotion that lacks reality or depth, falling flat in its attempts to depict real life and achieving only feminine melodrama' (Williamson, 2014a: 1). Such dismissals are at odds with the argument that sentimentalism has the capacity in art and literature to invite 'us sympathetically to share the emotional world of those distant from us in time and circumstance' (Brown, cited in Nyman, 2016: 67). In cinema, sentimentalism has been recovered through feminist critiques of the women's film and melodrama (Gledhill, 1987) and through a reappraisal of the postmodern sensual experience in contrast to the aesthetic disinterest of 'terminal irony' (Burnetts, 2017: 15). In television studies and elsewhere, sentimentalism has been proposed as an antidote to postmodern nihilism and as a rhetorical mode that makes socio-political analysis and complex identifications accessible to contemporary audiences (Williamson, 2014b: 11). Yet, when it comes to animals, pejorative alignments between empathy, anthropomorphism and sentimentality remain evident (Aaltola, 2012: 176).

One problem with this thinking is that it draws on long-held views that disavow the legitimacy of emotional responses towards animals. These criticisms have assumed that emotionality signals some sort of diminished capacity for reason. They rely on the normalisation of an emotion-reason binary, a result of which is that emotion is regarded with suspicion, as undisciplined, irrational, unpredictable and susceptible to manipulation. In this context, anthropomorphism is then posited as the mechanism by which that emotional manipulation can take place. The reductive assumption that is then allied to such thinking supposes that emotion cannot exist in conjunction with reason. Emotionality is then associated with vulnerability and weakness while the ability to regulate and control emotions is thought to be a marker of intellect.

These dualisms have gendered dynamics and the feminisation of emotion, as Sarah Ahmed points out, retains an 'association between passion and passivity. [. . .] Emotions are associated with women, who are represented as "closer" to nature, ruled by appetite, and less able to transcend the body through thought, will and judgement' (Ahmed, 2014: 3). Such associations were reflected in, for example, the nineteenth century when women's support for the antivivisection movement in Britain drew accusations that their opposition 'to rational science' was motivated by emotion, hysteria or sentimentality (Molloy, 2011: 28–29). In France and America, sentimentality, particularly for animals, was associated with a gendered irrationality which was, in some cases, medicalised and thought to be a sign of mental illness, while in other quarters it was proposed that women should be properly 'trained' to supress their emotions (Molloy, 2011: 29–31). Carol J. Adams notes that 'emotions are denigrated as untrustworthy and unreliable. They have long been viewed as invalid sources of knowledge. Moreover, they are equated with women, with being "womanish"' (Adams, 2007: 201). As

Adams and others makes clear, the connections between gender and emotion are not the relics of nineteenth-century discourses; the connotations continue to circulate in contemporary life (Williamson, 2014b). Yet, as Aaltola elegantly summarises, 'rationalism defends the notion that moral agency rests on the use of reason, sentimentalism maintains that the roots of such agency are to be found in emotions' (Aaltola, 2018: 2).

The ways in which emotion has been castigated can be seen to intersect with anxieties about maintaining a myth of evolutionary development to a state of human exceptionalism. Ahmed suggests that a Darwinian evolutionary discourse has reinforced the idea that emotions belong to primitive humans and function in narratives as 'a sign of "our" pre-history' (Ahmed, 2014: 3). The risks of 'regression' to a primitive state of being are built-in to narratives that activate connections between emotionality and an animal-like state. These narratives have longevity, and James R. Averill goes back to Plato to illustrate a lengthy historical legacy of 'extrinsic symbolism [which] has linked the emotions to the irrational, non-cognitive, primitive, animal-like and visceral' (Averill, 1996: 217). Here, we run into the odd contradictions of such connections where the relationship between emotions and being animal-like can exist beside 'rational' scientific arguments which insist that only humans have emotional experiences. To suggest otherwise is to anthropomorphise. At these junctures, far from being mutually exclusive, the symbolic and the scientific reinforce one another. Emotion is conceived of as animal-like, yet animals have historically been denied the capacity for emotion and in some spheres emotion still remains a specifically human characteristic.

Much crucial work has been done by feminist scholars to critically interrogate the pejorative associations between gender and emotion. For instance, in relation to how this plays out within science, Val Plumwood has offered a compelling account of the ways in which hierarchies of knowledge and their attendant practices reinforce the 'concept of scientific disengagement' that, she argues, 'is a powerful constituting and normative mythology for science, and perhaps, given the strong and continued gendering of reason/emotion dualisms in dominant global culture, the one that most strongly marks science out as a masculinist activity' (Plumwood, 2002: 42). Donna Haraway illustrated how the gendered reason/emotion dualism can manifest when she wrote about female primatologists who have 'deliberately taken advantage of the greater latitude for women in western culture to acknowledge emotional exchange with the animals and to affirm the importance of identification or empathy in a way that they believe improves research', but she notes that

> these same women, as well as many who allow no greater identification than their male peers, repeatedly report having to guard against incautious admission or cultivation of their feelings, in order to be respected scientifically or to avoid being labelled 'naturally' intuitive.
>
> (Haraway, 1989: 249)

In her interrogation of the gendering of scientific objectivity, Haraway exposes a crucial link between power and the apparent universality and dominance of scientific language (Haraway, 1988).

According to Plumwood, a commitment to scientific disengagement necessarily eliminates any relationships of care and sympathy to the extent that the object of knowledge becomes bound into inescapable instrumentalist relations of power where 'what is known becomes a means to the knower's ends' (2002: 42). Although there are exceptions, anthropomorphism generally has also been constructed in opposition to science's instrumentalist disengagement, strengthening the authority of science's knowledge claims to unbiased, value-free objectivity. In her critique of such moves, Plumwood argues that disengagement is part of an apparatus of power and wielded by the powerful 'to employ the well-practised conceptual and emotional distancing mechanisms which legitimate the exploitation of the objectified and oppressed' (2002: 44). Anthropomorphism is therefore entangled in relations of power where its exclusion and the discursive construction of pejorative associations with feminised emotion, 'primitive' thinking, childishness and popular knowledge can be deployed by a discourse of rational instrumentalism to justify exploitative animal practices and to discredit those who engage in empathetic concern for other beings.

These manoeuvres are challenged within the feminist ethic of care tradition, where emotion is reclaimed and regarded as crucial to the transformation of relationships between humans and other animals. Some feminist scholars of animal ethics have posited a place for emotional fellowship with other animals that diverges from rights theory, which has explicitly rejected emotion and sentimentalism. Josephine Donovan has argued that animal rights approaches which refuse emotion and are concerned about 'being branded sentimentalist are not accidental; rather they expose the inherent bias in contemporary animal rights theory toward rationalism, which paradoxically, in the form of Cartesian objectivism, established a major theoretical justification for animal abuse' (Donavan, 2007: 59). Because of this, the care tradition in animal ethics is often set in opposition to traditional ethical theories that rely on abstract principles and value impartial reasoning. This has resulted in an artificial division where justice and care ethics are mapped onto simple gendered associations with 'masculine' and 'feminine' characteristics. In her response to this binary, Lori Gruen argues that care is not a ' "feminine" theory or a "woman's ethic" ' (Gruen, 2015: 32). She outlines how care theory does provide an alternative to traditional approaches, noting the emphasis on context, relationality and connection as opposed to abstraction, individualism and impartiality. Gruen contends that justice and care are not incompatible binaries and she writes: 'Importantly, reason and emotion cannot be meaningfully separated either as they are mutually informing. Any compelling moral theory has to recognize that cognition/reason and affect/emotion cannot be disentangled' (Gruen, 2015: 34).

Donovan calls for a feminist ethic that has as its basis a relational culture of care and attention. She argues 'we should not kill, eat, torture or exploit animals because they do not want to be so treated, and we know that. If we listen, we can hear them' (Donovan, 2007: 76). According to Donovan, an ethic grounded in an emotional conversation with other species can contribute to transformative animal-human relationships. Donovan makes explicit the importance of attention – of being attentive to other beings – and Marti Kheel elaborates on how this relates to empathy when she writes 'Caring for other-than-human animals can only flourish with the aid of empathy. Empathy, in turn, can be seen as the culmination of many small acts of attention' (Kheel, 2008: 227). The type of care that both Kheel and Donovan are interested in is one which attends to individuals, is appropriate and contextualised, and which pays attention to particular situations. Being attentive then, in this context, is attention to difference and an acknowledgement that the category 'animal' is in many ways problematic because it cannot accommodate the immense variety of individual embodied experiences, behaviours, habitats and so forth that exist. Kheel writes, 'cumulatively, these acts of attending can help us to appreciate other-than-human animals as individual beings with subjective identities, rather than merely part of a larger backdrop called the "biotic community", "the ecosystem", or "the land" ' (ibid.).

The language used in ethic of care scholarship has meant that anthropomorphism as a concept and practice has had to be addressed head-on. For example, Kheel argues that an ecofeminist philosophy 'affirms the integrity of individual other-than-human animals both domestic and wild. It begins with the simple observation that other animals are individual beings with feelings, needs and desires' (Kheel, 2008: 227). Plumwood sets out the direction of travel for the debate when she writes, 'there is no good logical reason why we should not speak of the nonhuman sphere in intentional and mentalistic terms, as we do constantly in everyday parlance, and would hardly be able to avoid' (Plumwood, 2002: 56). She proceeds to argue that 'there is no basis for the general claim that speech is invalidated by anthropomorphism merely on the grounds that it attributes intentionality, subjectivity or communicativity to non-humans' (ibid.). Unravelling the nebulous definitions of anthropomorphism, Plumwood argues that there are two senses of the term: the attribution of human characteristics to nonhumans, and the attribution to nonhumans of uniquely human characteristics. Both, Plumwood proposes, are flawed. The first assumes that there are no shared characteristics – what she refers to as 'hyper-separation' of human and animal natures – and the second assumes 'that non-humans do not have characteristics such as subjectivity and intentionality' (Plumwood, 2002: 57). In these two senses, Plumwood suggests, what is usually referred to as anthropomorphism is instead anthropocentrism.

Attending to the relationship between empathy and anthropomorphism, Lori Gruen suggests that certain kinds of empathising can lead to 'problematic anthropomorphizing', where another being's situated position is not taken into account (Gruen, 2015: 66). In this case, difference is erased and only similitude is acknowledged. To avoid this anthropocentric empathy and the resultant anthropomorphism that might have deleterious effects on nonhuman animals, Gruen argues for 'entangled empathy' which 'involves a particular blend of affect and cognition' (ibid.). Alternating between first- and third-person points of view, Gruen proposes 'the empathizer is always attentive to both similarities and differences between herself and her situation and that of the fellow creature with whom she is empathizing' (ibid.). Gruen's entangled empathy thesis thus finds a space for anthropomorphism to be productive in the oscillation between viewpoints and the marshalling of both cognition and affect. It is also of particular interest to this book as it clearly has implications for cinema, media and the moving image generally, where spectatorial identification and viewpoint have been widely theorised since the 1970s. Moreover, phenomenological approaches to cinema that interrogate the embodied experience of the film viewer draw our attention to the sensual experience of 'watching' film and other media. In this regard, identifications between the 'viewer' and a nonhuman animal onscreen can move beyond the oculocentric position that has informed much film theory and embrace the multisensorial experience of media engagement by which another subjectivity can be understood affectively (Parkinson, 2018).

My focus on empathy should not be taken to diminish the importance of compassion to ethical animal-human relations and advocacy. I am mindful of Lori Gruen's point that the terms 'empathy' and 'compassion' may be, to some degree, interchangeable, depending on the disciplinary background of the writer. However, empathy is sometimes set in contrast with compassion in debates about emotion and moral decision-making and while this book is not about empathy per se, it is nonetheless important to outline here why I have opted to use the term 'empathetic' rather than 'compassionate' connections. The reason is rather straightforward, in that a useful distinction between empathy and compassion has it that the former involves an imaginative reconstruction of another's experience while the latter requires there to be some degree of suffering (see, for example, Nussbaum, 2008). In other words, empathy can accommodate the imagined appreciation of another's joy, happiness or contentment as well as suffering, while compassion is particularly concerned with a judgement of distress. In terms of advocacy there is, without doubt, an argument to be made that the suffering of other species should be a motivation for action. Images that depict suffering might indeed move us, but then so too do images of nonhuman animal joy. In other words, *each* have affective dimensions. Compassion might be motivated by suffering, but empathy can imagine both painful and pleasurable experiences. In terms of the topic of book, emphasis on compassion would mean limiting the study to representations of suffering and, while the vulnerability of nonhuman animal bodies is indeed a catalysing force, the mobilising capacities of anthropomorphism are not confined solely to the empathetic imaginings of nonhuman animal distress. This does not mean of course, that images of nonhuman animal joy, for example, are benign or untethered from the material realities of those animals' lives. In the case of 'live action' or representations that use real individuals as a referent (for example the use of real tigers as 'reference' for artists and animators working on *Life of Pi* (2012)), it is important to consider the experiences of those individual nonhuman animals who have been filmed, photographed, drawn or otherwise recorded. Additionally, Aaltola makes a vital point in her comprehensive account of varieties of empathy when she refers to 'bite-size' forms of empathy that are delivered up by entertainment media and which allow human viewers to 'love' animals but not be mobilised into any meaningful action to reduce suffering (Aaltola, 2018: 44–45). I address this issue in my discussion of sites of commodified anthropomorphism and affective labour, where I consider the relationships between the material realities of nonhuman animals' lives and experiences and the affective engagements with their mediated anthropomorphised representations.

Summary

In the chapters that follow, the mediation of nonhuman animal life and experience is considered through a series of case studies in Chapters 3 through 6. Mediation is taken to be a confluence of the social, institutional, political, technological and embodied which connects the discursive and the affective, and it is the mediation of nonhuman animals that gives rise to anthropomorphism. To rethink it through this lens, anthropomorphism is proposed here as differentiated, situational, contextual and entangled. In terms of differentiation, I propose four sites of anthropomorphism: empathetic, affective animal labour, pejorative and commodified. Empathetic sites are those where anthropomorphism is an intersubjective imagining of another animal's experience. Affective animal labour sites are those where anthropomorphism is utilised for its affective power. In this sense anthropomorphism is a product of the affective labour of other animals, used as an anthropocentric salve. Pejorative sites are those where anthropomorphism is produced as a denigrated object of discourse, used to reclaim the authority of anthropocentric knowledge about other species. Sites of commodification are those that are in service to capitalism where anthropomorphism is appropriated as a strategy to engage humans as customers with the 'product' rather than the animal themselves. These are not types of anthropomorphism but instead sites that are shaped by context. They can and certainly do overlap and are therefore not mutually exclusive. Indeed, it is the messiness of their entanglements that are, I contend, the points where anthropomorphism may be productive as a catalyst for empathetic connections between humans and other animals.

Notes

1 In keeping with a critical animal studies approach I use the term 'nonhuman animal' or 'other animal' where the word 'animal' is not defined by context.
2 The scope of this book does not allow me to explore anthropomorphism in non-western cultural contexts, a limitation that I hope will be addressed in future scholarly work on the subject.

References

Aaltola, E. (2012) *Animal Suffering: Philosophy and Culture*, Palgrave MacMillan, Basingstoke.
Aaltola, E. (2018) *Varieties of Empathy: Moral Psychology and Animal Ethics*, Rowman & Littlefield, London and New York.
Adams, C. J. (ed) (2007) *The Feminist Care Tradition in Animal Ethics*, Columbia University Press, New York.

Ahmed, S. (2014) *The Cultural Politics of Emotion* (Second edition), Edinburgh University Press, Edinburgh.

Almiron, N. and Cole, M. (2016) 'Introduction: The convergence of two critical approaches' in Almiron, N. and Cole, M. (eds) *Critical Animal and Media Studies: Communication for Nonhuman Animal Advocacy*, Routledge, London, pp. 1–10.

Ang, I. (1996) 'Ethnography and radical contextualisation in audience studies' in Hay, J., Grossberg, L., and Wartella, E. (eds) *The Audience and Its Landscape*, Westview, Colorado, pp. 247–262.

Averill, J. R. (1996) 'An analysis of psychophysiological symbolism and its influence on theories of emotion' in Harr, R. and Parrott, W. G. (eds) *The Emotions: Social, Cultural and Biological Dimensions*, Routledge, London and New York, pp. 204–228.

Blackman, L. and Venn, C. (2010) 'Affect' in *Body and Society*, Vol. 16 (1), pp. 7–28.

Burnetts, C. (2017) *Improving Passions: Sentimental Aesthetics and American Film*, Edinburgh University Press, Edinburgh.

Calarco, M. (2008) *Zoographies: The Question of the Animal from Heidegger to Derrida*, Columbia University Press, New York.

Donavan, J. (2007) [1990] 'Animal rights and feminist theory' in Donovan, J. and Adams, C. J. (eds) *The Feminist Care Tradition in Animal Ethics*, Columbia University Press, New York, pp. 58–86.

Fairclough, N. (1992) *Discourse and Social Change*, Polity Press, Cambridge.

Foucault, M. (1997) *The Order of Things: An Archaeology of the Human Sciences*, Routledge, London.

Gledhill, C. (1987) *Home Is Where the Heart Is: Studies in Melodrama and the Woman's Film*, BFI, London.

Gruen, L. (2015) *Entangled Empathy: An Alternative Ethic for our Relationships with Animals*, Lantern Books, New York.

Haraway, D. (1988) 'Situated knowledges: The science question in feminism and the privilege of partial perspective' in *Feminist Studies*, Vol. 14 (3), pp. 575–599.

Haraway, D. (1989) *Primate Vision: Gender, Race and Nature in the World of Modern Sciences*, Routledge, New York and London.

Haraway, D. (1991) *Simians, Cyborgs, and Women: The Reinvention of Nature*, Routledge, London and New York.

Kheel, M. (2008) *Nature Ethics: An Ecofeminist Perspective*, Rowman & Littlefield Publishers Inc., New York.

Knudsen, B. T. and Stage, C. (eds) (2015) *Global Media, Biopolitics, and Affect*, Routledge, London and New York.

Merskin, D. (2016) 'Media theories and the crossroads of critical animal and media studies' in Almiron, N. and Cole, M. (eds) *Critical Animal and Media Studies: Communication for Nonhuman Animal Advocacy*, Routledge, London, pp. 11–25.

Molloy, C. (2001) 'Marking Territories' in *Limen: Journal for Theory and Practice of Liminal Phenomena*, Vol. 1, online at http://limen.mi2.hr/limen1-2001/clair_molloy.html

Molloy, C. (2011) *Popular Media and Animals*, Palgrave Macmillan, Basingstoke.

Niemyjska, A., Cantarero, K., Byrka, K., and Bilewicz, M. (2018) 'Too humanlike to increase my appetite: Disposition to anthropomorphize animals relates to decreased meat consumption through empathic concern' in *Appetite*, Vol. 127, pp. 21–27.

Nimmo, R. (2016) 'From over the horizon: Animal alterity and liminal intimacy beyond the anthropomorphic embrace' in *Otherness*, Vol. 5 (2), September, pp. 13–45.

Nussbaum, M. C. (2008) *Upheavals of Thought: The Intelligence of Emotions*, Cambridge University Press, Cambridge and New York.

Nyman, J. (2016) 'Re-reading sentimentalism in Anna Sewell's *black beauty:* Affect, performativity, and hybrid spaces' in Nyman, J. and Schuurman, N. (eds) *Affect, Space and Animals*, Routledge, London and New York, pp. 65–79.

Nyman, J. and Schuurman, N. (2016) 'Introduction' in *Affect, Space and Animals*, Routledge, London and New York, pp. 1–9.

Parkinson, C. (2018) 'Animal bodies and embodied visuality' in *Antennae: Journal of Nature in Culture*, Vol. 46, pp. 51–64.

Plumwood, V. (2002) *Environmental Culture: The Ecological Crisis of Reason*, Routledge, New York.

Ryder, R. D. (1989) *Animal Revolution: Changing Attitudes Towards Speciesism*, Blackwell, London.

Sobchack, V. (2004) *Carnal Thoughts: Embodiment and Moving Image Culture*, University of California Press, Oakland.

Spitulnik, D. (2010) 'Think context, deep epistemology: A meditation on wide-angle lenses on media, knowledge production and the concept of culture' in Brauchler, B. and Postill, J. (eds) *Theorising Media and Practice*, Berghahn Books, Oxford and New York, pp. 105–126.

Tam, K. P. (2015) 'Mind attribution to nature and proenvironmental behavior' in *Ecopsychology*, Vol. 7, pp. 87–95.

Taylor, N. (2011) 'Anthropomorphism and the animal subject' in Boddice, R. (ed) *Anthropocentrism: Humans, Animal, Environments*, Brill, London, pp. 265–282.

Taylor, N. and Twine, R. (eds) (2015) *The Rise of Critical Animal Studies: From the Margins to the Centre*, Routledge, London and New York.

Waytz, A., Cacioppo, J., and Epley, N. (2010) 'Who sees human? The stability and importance of individual differences in anthropomorphism' in *Perspectives on Psychological Science*, Vol. 5, pp. 219–232.

Weil, K. (2012) *Thinking Animals: Why Animal Studies Now?* Columbia University Press, New York.

Williamson, J. (2014a) 'Introduction: American sentimentalism from the nineteenth century to the present' in Williamson, J. and Larson, J. (eds) *The Sentimental Mode: Essays in Literature, Film and Television*, MacFarland and Company, Inc., North Carolina, pp. 3–14.

Williamson, J. (2014b) *Twentieth-Century Sentimentalism: Narrative Appropriation in American Literature*, Rutgers University Press, Brunswick, NJ and London.

2 Anthropomorphism, mediation and authority

Introduction

Anthropomorphism is tricky. In terms of an interspecies dynamic it can be understood as a projection of human characteristics onto other animals, practices that bring animals into the human world, or a relational state that enables intersubjectivity between human and nonhuman animals. Each of these ways of thinking about anthropomorphism implies a different power relation and a possibility or not for empathetic connections. As such, anthropomorphism weaves a complicated path through discourses of science and popular culture, the particularities of its history as a discursive object emerging through retrospective evaluations of its efficacy as method, its alignment with certain (denigrated) human traits and the role it plays in category confusion. The projection of whatever constitutes the 'properly human' onto whatever or whoever falls outside of that narrow classification, whether that be deities, objects or other animals, has triggered all manner of ontological anxieties and moves to repair the epistemological cracks that have allowed such leakages. Messy and contentious, anthropomorphism has its own discursive history, and it is through the criticisms levelled at it that we arrive at an understanding of what it means as a practice at any given time. In this chapter, I discuss the four sites of anthropomorphism identified and outlined in Chapter 1 in relation to the mediation of animal lives and experience. Informing this chapter is the question of power and 'who' is culturally authorised to mediate.

Pejorative sites

The denigration of anthropomorphism has a rich history, one that is usually traced back to Xenophanes' critique of the human impetus to create God in the image of 'man':

> if oxen and horses and lions had hands and were able to draw with their hands and do the same things as men, horses would draw the shapes of gods to look like horses and oxen to look like oxen, and each would make the gods' bodies have the same shape as they themselves had.
>
> (Xenophanes, in Fieser, 2001: 3)

While it is outside the scope of this discussion to deal with anthropomorphism and religion, I will wrestle this quote away from its historical context because, decontextualised, it is riven through with many prevailing prejudices that continue to inform human-animal relations long after Xenophanes' demise. If the idea of horses and oxen engaged in creative production remains amusing to our contemporary sensibilities, does that humour derive from the normative denial of agency, autonomy, creativity and culture to animals and the notion of 'lack' that has traditionally propped up the anthropocentric claims that animals are inferior to humans? Of course, if species inferiority is always measured against human standards, then in setting the rules of domination, humans normalise the exploitation of other sentient beings. In this regard, we can understand one 'problem' of anthropomorphism emerging from intellectual efforts to shore up the distinctions between subject and object, a move which reassures human exceptionalism and supposes humans to occupy a position apart from the world.

Another power dynamic suggested by anthropomorphism is that species difference is at best discounted and at worst erased, to leave us with quasi-human analogues in animal skins, their value residing in a capacity to entertain or fulfil the craving for cute otherness. Anthropomorphism conceived of as narcissistic desire is interminably tied to anthropocentric conceit or some naïve misunderstanding of difference, and it is this critique of the humanisation of animals that is often readily directed towards media and popular culture. However, the animal-shaped human proxies that populate our screens also force us to confront some difficult questions about their affective appeals and mobilising potential. Somewhere between the ideal of interspecies intersubjectivity, a pragmatic desire for greater public empathy for other animals and the reality that anthropomorphic practices are well entrenched within everyday life lies a tension. First, there is a strong impetus from various intellectual traditions to critique naïve anthropomorphism as human-centred and narcissistic which dismisses anthropomorphic affect and its potential to mobilise empathy. However, as Daston and Mitman suggest, the circulation of animal images has widened human sympathies and attuned public understanding to the extent that 'anthropomorphized expressions of other animals may now be viewed as more humanely intelligible than those of other humans' (2005: 5). In other words, while we might ache with discomfort and bemoan the public desire for charismatic megafauna, cute cat videos and humans voicing the imagined interior monologues of other species, such representations have been active in raising public awareness and concern for other animals. Moreover, in expressing our critical discomfort about, for example, the infantalisation of dogs, are we indulging in a problematic projection of our own sensibilities about animal dignity that takes no account of the reality of the dog's experience? In our critiques of anthropomorphism, where does another form of anthropocentric projection take over? Second, what are the terms of a critique of anthropomorphism where it arises within media and popular culture? Media texts are polysemic, they do not have singular effects and they are not consumed in a social or cultural vacuum. For this reason, the reception of anthropomorphism in media texts is, I contend, better understood as situational and contextual rather than as a singular problem of anthropocentric vanity or charged with the offence of inaccuracy or categorical error. It is surely problematic to bring the delimitations and scepticism of scientific discourses to bear on popular forms of anthropomorphism even where it is applied critically in a methodological sense to the study of other animals. Other than perhaps a small canon of critically 'acceptable' texts, popular media will always be found wanting if those criteria are used and such a framework does not account for the specificities of cultural production, where live animals figure in that process, or how popular anthropomorphism shapes our encounters with other animals and public sensibilities about animal issues.

Pat Brereton (2016) suggests that within contemporary culture most people acquire their ethical attitudes from engaging with mass media. In relation to eco-cinema, Brereton identifies a similar problem with critical responses to commercial film as I suggest happens all too readily with anthropomorphism in popular media. Some scholars argue that only certain types of 'independent lyrical and activist documentaries might be thought of as eco-cinema', but, Brereton contends,

> *all* types of film, from the excesses of a commercial Hollywood blockbuster, alongside the more rarefied and explicitly ecological art-house narrative, can [. . .] highlight specific ecological issues and ethical debates and thereby situate these concerns within the general public consciousness.
>
> (Brereton, 2016: 31)

In the case of anthropomorphism, we should not discount commercial entertainment for crass simplification or think that it is unable to engage an audience in an ethical dialogue about other species. Like Brereton, I'm inclined to consider the commercial, the popular, the mainstream; in this case because I contend that these sites of anthropomorphism can be *effectively* affective.

In the background of critiques over legitimate eco-cinema and popular anthropomorphism sits the normalisation of scientific scepticism and the authority that is culturally ascribed to scientific discourse. The public reception of commercial films that produce, consciously or inadvertently, a critical commentary on climate change for example, will usually make space for the scientist's view. It is customary, especially for commercial films that have resonated with audiences, to see press articles that ask for an evaluation of 'the science'. So, we find, for instance, headlines such as 'A palaeontologist's view of *Ice Age; Continental Drift*' (Barnett, 2012) or 'Ice age movie is realistic, says Britain's chief scientist' (Connor, 2004) after the releases of the fourth *Ice Age* film and Roland Emmerich's *The Day After Tomorrow*. The push and pull for legitimacy that is played out in the public domain as a good-natured swipe, a wholesale discrediting or (partial) endorsement of the creative imaginings of commercial media is a familiar discursive tussle. It is especially important in the case of anthropomorphism where I contend that the contest between human interpretations of animal life and experience can invoke scientific discourse as a normalised objective authority in opposition to an intersubjective empathetic position. While it is not always 'science' that is positioned as the great leveller and arbiter of authoritative knowledge, as the preceding climate change examples illustrate, there is a tendency for public discourse to set science and the popular imaginary in some sort of antagonistic tension. It is important therefore to examine the contact points between science and the popular as these are some of the most contentious sites of anthropomorphic practice. In other words, this set of conditions can give rise to the sites where anthropomorphism is produced in a pejorative sense, as a denigrated object of discourse.

The discomfort of anthropomorphism

Whether critical of anthropocentric thinking or aligned with humanist views, anthropomorphism continues to cause discomfort. If its history suggests anything, it is that the anthropomorphism we have reflects the way in which we understand ourselves and what it means to be human at any given time. In this sense, the anthropomorphism of Victorian Britain reveals much about the human social hierarchies and sensibilities of the time and contemporary anthropomorphism equally reflects current social and cultural norms and attitudes. Thinking of anthropomorphism in this way reveals something of human social structures and norms, but it does not necessarily give much insight into human-animal relationships. Instead, we can consider anthropomorphism as an object of discourse which has relied on the notion that humans are, in some respect or other, different from other species. Crucially, it is how that difference is articulated and discursively deployed that is important in terms of how interspecies asymmetries of power are linked to anthropomorphic practices. Language, morality, emotion, consciousness, reason, and 'a soul' have, at different points in history, been naturalised as characteristics that are uniquely attributable to humans. These differences have then been crucial to the maintenance of ideological boundaries between human and other animals. In turn, those boundaries have then become central to the identification and denigration of anthropomorphic practice, where it is considered an error to attribute other animals with characteristics reserved solely for humans. Prior to the late nineteenth century anthropomorphism referred to the attribution of human form or characteristics to deities. By the early twentieth century anthropomorphism was commonly used to refer to the humanisation of animals, and Charles Darwin's work was important to that redefinition. As one encyclopaedia declared in 1909:

> The substitution of Darwin for Paley as the chief interpreter of the order of nature is currently regarded as the displacement of an anthropomorphic view by a purely scientific one: a little reflection, however, will show that what has actually happened has been merely the replacement of the anthropomorphism of the eighteenth century by that of the nineteenth.
>
> (Geddes, in Seward, 1909: 49)

Darwin's work was not the reason that anthropomorphism acquired a pejorative status in the twentieth century, but it became representative of what was retrospectively defined in some spheres as anthropomorphic.

A fulcrum in the shift from nineteenth-century linguistic, sentimental and cultural 'excesses' to their twentieth-century rejection, Richard Ryder (1989) suggests that the ethical implications of Darwin's human-animal continuum were potentially perilous for anthropocentrism. However, Ryder argues, the dissemination of knowledge about evolutionary kinship within British Victorian society was not pivotal to the advancement of humane attitudes towards animals at the time. Instead, he suggests that social stability and a fashionable preoccupation with pain and suffering were the greater catalysts in mobilising popular support for animal welfare. It was not until the following century, Ryder argues, that animal liberation became concerned with the moral implications of evolution. Historian Harriet Ritvo argues that Darwin's continuity thesis, while transformative in principle, did not trouble the popular discourse of zoology which was widely consumed by a multigenerational audience and had already established a place for humankind amongst the animals (1987: 31–41). And, in an historical study of Victorian attitudes towards animals and pain, James Turner similarly details how the animal protection movement at the time was ambivalent towards Darwin's work, evolutionary theory having the benefit of an extant understanding of human-animal biological similitude that had developed over the previous hundred or so years (1980: 60–61). As these studies suggest, the alignment of anthropomorphism with the mobilisation of public attitudes towards animal welfare does not have a straightforward correlation with the Darwinian human-animal continuum.

However, the Darwinian discourse does become associated with the later pejorative understanding of anthropomorphism and Darwin is a useful point of entry into the contest over the legitimacy of anthropomorphism as method, a mode of interpretation, and an inevitable outcome of mediation and popular discourse. Darwin's theory of natural selection granted nature agency, his concept of species continuity blurred definitions of the 'properly human' and challenged prevailing ideas about the boundary between 'man' and 'beast'. However, the concept of species continuity was not, in itself, anthropomorphic. Rather, anthropomorphism and aspects of species continuity became entwined as the boundaries between human and animal were reshaped by new claims to knowledge and the use of anecdote and popular knowledge were criticised as methodologically problematic. The Romantic Movement in British literature had been key to the development of a discourse in which nature assumed a powerful agency and fleeting sublime beauty. Darwin's authorial style maintained identifiable connections with the sensibilities of Romanticism through the attribution of agency to nature and the elegant descriptions of its minutiae (Beer, 1984; Turner, 1980). Darwin's written descriptions of animal behaviour were thus retrospectively branded anthropomorphic by twentieth-century behavioural scientists; his writing judged to be a peculiar stylistic relic of the previous century.

Distinctions between humans and nature had been established within the religious and scientific knowledge in the seventeenth and eighteenth century and these discourses remained, to a large extent, interconnected until the late nineteenth century. At the end of the nineteenth century science became professionalised and institutionalised, a move that established distance from religious discourse and was concurrent with a rejection of anthropomorphic practices. By the beginning of the twentieth century, practices that were thought to humanise nonhuman animals were largely censured within science and serious art (Baker, 2000) but continued to be embraced by popular culture. New divisions were established between 'serious' and 'popular' ways of thinking and talking about animals, and this demarcation located anthropomorphism firmly within the sphere of cultural fictions. Early twentieth-century western science actively avoided attributing nonhuman animals with what were classified as uniquely human characteristics while serious art and literature distanced themselves from the sentimental anthropomorphism of the previous century. A rejection of anthropomorphic practice was linked to notions of modernism, professionalism and credibility. Allied with this, the denigration of sentimentalism and the buttressing of anthropocentric ideas of difference were conflated with the idea that human intellectual maturity was signalled by a move away from the anthropomorphic tendency and toward a rational scientific understanding of the world.

Emergent methods of scientific enquiry in the transition from the nineteenth to the twentieth century dislodged to a large extent the legitimacy previously ascribed to mental and emotional continuity between humans and other animals. The history of this turn is well documented, but it is useful to recount here in brief a narrative of the shift. In 1903, the Russian physiologist Ivan Pavlov presented his study, 'The Experimental Psychology and Psychopathology of Animals', at a medical conference in Madrid, on the conditioned reflex in dogs (Pavlov, 1963). Pavlov's study challenged Darwin's theory of innate modes of expression when he showed that the salivation response in a dog could be stimulated by external means and was therefore a learned response. Four years later, in 1907, psychologist Oskar Pfungst published his study *Das Pferd des Herrn Von Osten (Der Kluge Hans). Ein Beitrag zur experimentellen Tier- und Menschen-Psychologie* ('The Horse of Mr Von Osten (Clever Hans): A Contribution to Experimental Animal and Human Psychology') (Pfungst, [1907] 1998). Pfungst refuted claims by Prussian aristocrat Wilhelm Von Osten that a horse named Clever Hans could perform mental arithmetic and showed instead that he was responding to subtle visual cues from the trainer rather than doing the calculations. In 1911 American comparative psychologist Edward Thorndike published *Animal Intelligence*. In this text Thorndike refuted all claims that animals possessed reason and claimed that in the extant literature 'most of the books do not give us a psychology, but rather a *eulogy* of animals. They have all been about animal *intelligence*, never about animal *stupidity*' (Thorndike, 1911: 22). Thorndike argued that rather than observe animals in nature, it was necessary to construct laboratory experiments to test hypotheses about intelligence. To these ends, he constructed a series of experiments on cats and dogs and argued that the 'experimental subjects' only showed evidence of learned responses, often generated from accidental actions. In 1913, the founder of the behaviourist school, J.B. Watson, wrote: 'The time seems to have come when psychology must discard all reference to consciousness; when it need no longer delude itself into thinking that it is making mental states the object of observation' (Watson, 1913: 161). Watson voiced his concern that psychology could not attain the status of a science until introspection, consciousness, mental states and mind were expelled from the discipline. He demanded that the only way forward for psychology was through the objective study of behaviour without recourse to internal states of mind or analogy through introspection. In his article 'Psychology as the Behaviorist Views It', Watson (1913) denied the possibility of mental continuity between humans and animals on the basis that mental states were in themselves not an object of study (Watson, 1913: 158).

Watson's claims were indicative of a new wave of reductionism that stretched from the influence of the behaviourist school within psychology and across scientific discourses. As psychology refuted the study of consciousness, internal states and introspection, there was a corresponding shift towards the idea that the language of science could be purified of extraneous linguistic frivolities if it was reduced to propositions that could be determined through empirical data as either true or false (Wittgenstein, [1921] 2001). This shift formalised the rejection of anecdotal evidence that had been the basis of Darwin's arguments and demanded that what was said or written about animals was purged of all subjective interpretation. This suggested that reality could be accessed via language, and it was therefore through language that the categorical error could be made. In short, the attribution of particular characteristics from one category could be mistakenly applied, through language, to another category. Consequently, during the first quarter of the twentieth century there was a major shift in the way in which scientific discourse could talk about nonhuman animals. Moreover, the regulation of discourse made explicit not only what could and what could not be said about animals, but also which questions could be asked in the first place.

Highly influential in the rejection of interpretative methods, psychologist C. L. Morgan argued that an emotional and mental continuum was inferred from the naturalist's own subjective state of mind. Morgan contended that 'the difficulty is due to the fact, that the only mind with which we can claim any first-hand acquaintance is the civilised mind, that of which we are conscious within ourselves' (Morgan, 1903: 42). Morgan argued that any inference of state of mind from the subjective standpoint of the observer had to be questioned. Morgan subscribed to the anthropocentric character of humanist science when he contended that the study of 'other' minds should be dismissed in favour of the scientific examination of the human mind. In opposition to the methodologies that Darwin's colleague George Romanes had utilised, Morgan made apparent the distinction between the professional scientist and the 'amateur naturalist' and the different methods of knowledge production accorded to each: the study of objective manifestations, the use of laboratory methods and dissection, introspection versus objective study and the overall endeavour of the professional scientist to systematise knowledge.

Despite later claims that it had been misrepresented as a principle of parsimony (Greenberg and Haraway, 1998), Morgan's argument rendered problematic the popularisation of science where naturalists' accounts of nonhuman animal life utilised the practices that Morgan deemed to be opposed to 'professional science'. Contemporaneous with Morgan's argument and methodologically indebted to

Darwin, naturalist accounts of nonhuman animal activity, in their own habitats, continued to be popularised within mass-produced periodicals for the educated middle classes. Morgan's argument made clear that the discourse of comparative psychology was differentiated from that of the naturalist by recourse to a distinction between amateur and professional practices. At the heart of this distinction were interpretative practices which crucially led Morgan to conclude that 'in no case is an animal activity to be interpreted in terms of higher psychological processes' (Morgan, 1903: 59). This principle was referred to as Morgan's Canon and, akin to Occam's razor, it was widely equated with the notion that all animal activity should be interpreted at the most basic, objective and verifiable level. In other words, rather than attribute mental states to an animal, scientists should explain activity in behavioural terms and without recourse to subjective interpretation. Morgan's contemporary, George Herbert Mead, likewise argued against subjective interpretation and claimed it was only suitable for poets and artists; it was instead, he argued, truth and facts that occupied the scientist (Mead, [1907] 1964: 73–81).

It is worth noting that the rejection of anthropomorphism from 'serious' science, and for that matter 'serious' art and literature, happens as cinema is established as a dominant mass media form in the early decades of the twentieth century. Indeed, as film language concretised its conventions through shot types, camera movement, transitions and continuity editing, scientific discourse sought to rid itself of the subjective anthropomorphic excesses that had characterised earlier scientific language and method, and modernist literature eschewed the sentimental mode. Nonhuman animals were central to the development of early cinema, their representations on screen organised through audiovisual systems that were at every level structured around human aesthetics and senses. Shot types referred to the human body for their organisation of scale; continuity editing conventions were based on human notions of time and space and patterns of movement and communication. In other words, the conventions of western cinema and logical positivism in the sciences both developed within a humanist paradigm in the first decades of the twentieth century. At the same time there emerged new knowledge formations that endorsed discursive distinctions between the artist and the scientist, between subjective and objective, between interpretation and fact. This epistemological legacy has continued to follow anthropomorphism into the contemporary era.

By the mid-twentieth century mentalistic descriptions of nonhuman animals were avoided and anthropomorphic thinking was widely considered incompatible with the scientific study of behaviour (Kennedy, 1992). Towards the end of the twentieth century a tolerance for certain anthropomorphic terminology in addition to an endorsement of anthropomorphism in a methodological sense gained a degree of legitimacy in some quarters (Sealey and Oakley, 2013: 401–402). However, while there have been shifts in scientific knowledge about animals, their emotions and behaviours, when it comes to popular anthropomorphism, the post-Darwinian discourse that shaped the interpretation of animal life and experience continues to echo loudly. It acts as a regulatory force, particularly at the contact points between science and popular culture where there remains a contest for authority over who can make legitimate claims about the lives and experiences of non-human animals. We should ask, then, who emerges from this very human epistemological battleground as an authoritative mediator?

Mediating animal experience

Using the term 'weak anthropomorphism', Val Plumwood refers implicitly to the processes of mediation when she argues that all human-to-human communication, and especially popular culture, inevitably anthropomorphises. She writes, 'any representation of speech-content for a human audience will have to be an interpretation in terms of human concepts and in that weak sense a background level of anthropomorphism is always likely to be present' (Plumwood, 2002: 57). She points out the problems of representation wherein the audience for the representation will dictate the level of anthropomorphism present and she concludes that 'the charge of anthropomorphism is a pseudo-scientific, rationalist convention which tried to reinforce human/nature dualism through an enforced conformity to hyper-separated vocabularies' (Plumwood, 2002: 61). In a comprehensive study of anthropomorphism and behavioural science literature, Eileen Crist (1999) explores this issue of scientific vocabularies and argues that the language used to depict nonhuman animals is never neutral. Highlighting the differences between ordinary and technical language, Crist notes that the former 'reflects a regard for animals as acting *subjects*' while the latter 'paves the way toward conceptualizing animals as natural *objects*' (Crist, 1999: 2, italics in original). Although the two domains of language are not exclusive, offering the sharp distinction and focusing on the work of language, Crist argues, makes apparent 'that similar behaviors can be represented in ways that produce extremely different effects on the reader's understanding' (ibid.). Language is, Crist proposes, 'a medium of travel for the reader to look upon animal life' (Crist, 1999: 3). Crist makes clear the point that mediation which broadly conforms to ordinary language – 'the everyday language of human action' – will inevitably confer meaningfulness, authorship and continuity to nonhuman animal action and in doing so construct a 'gateway to a landscape that has the forms and ambiance of subjectivity' (Crist, 1999: 4–5). This language produces anthropomorphic depictions that align human and nonhuman animal worlds whilst behaviour mediated by technical language produces mechanomorphic depictions that maintain distance between those worlds. In mechanomorphic depictions 'animals become portrayed as vessels steered by forces they neither control nor comprehend, and as a consequence, behaviors emerge as events that happen to animals, rather than as active achievements' (Crist, 1999: 203). Although, as Crist rightly points out, there is crossover between the ordinary/ technical, subject(ive)/object(ive), anthropomorphic/mechanomorphic and relational/distanced, these dualisms do summarise well the struggles faced in representing nonhuman animals.

The realities of such struggles are well documented in ethology, where scientists such as Mark Bekoff have given personal accounts of their experiences. Bekoff writes, 'scientists are taught not to empathize, but rather to remain detached and objective. Naming animals and assigning them personalities were widely frowned upon as being "too anthropomorphic" and non-scientific. But, this is where science has gone wrong' (Bekoff in Gruen, 2015: vii). Mary Midgley also sees science as having 'gone wrong' in this regard and she asserts that anthropomorphism is a confused concept and the only example where a notion 'invented solely for God' is 'then transferred unchanged to refer to animals' (Midgley, 1998: 125). For Midgley, evolution makes it implausible that animals should be considered 'so unlike us' and she concludes that in having powers of expression and sympathy humans can, through care and attention to their interactions with other animals, identify their moods and feelings (Midgley, 1998: 128–133). Midgley proceeds to take the behaviourist's principle of parsimony and turn it around to argue that attempts to describe behaviour in external terms result in a convoluted discourse that does not open up easily to further questioning. Instead she maintains, attributing subjective states to animals provides a much simpler explanation and enables the observer to go on to make further useful predictions (1998: 133–135). Bekoff follows in the ethologist's tradition that Midgley observes has revolutionised the ways in which animal motivation and communication has been studied. Bekoff gives credence to Midgley's argument in support of anthropomorphism when he states that 'hard science is confirming what our intuitions so often tell us: animals express emotions in ways we are naturally able to understand' (Bekoff, 2007: 53).

Bekoff and others make a case for anthropomorphism and argue that far from being a hindrance, anthropomorphism confers advantages to scientists and can be instrumental in the furtherance of scientific knowledge about other species (Law and Lien, 2017). This remains contentious, and there are still many who continue to reject anthropomorphism as unscientific and lacking objectivity (Kennedy, 1992; Wynne, 2007; Dawkins, 2012). Attempts to negotiate the validity of what has been referred to as an 'innate' or 'inevitable' human trait (Kennedy, 1992; Guthrie, 1993) have, since the late twentieth century, led to the categorisation of anthropomorphism into different types, sometimes hierarchised to identify its censured and accepted forms. These typologies have been proposed as useful for thinking through the 'problem' of anthropomorphism, a mechanism by which it can be categorised, named, and a corresponding value judgement applied. We can understand the development of typologies of anthropomorphism as indicative of motivations to authorise, or not, its methodological legitimacy within particular disciplinary fields of enquiry. It is perhaps then unsurprising that classificatory endeavours have come mainly from the sciences and philosophy.

In the case of the sciences, a number of prominent primatologists and ethologists have spoken out about their use of anthropomorphism and the professional challenges that scientists face in having such practices recognised as 'good' science. For instance, in *The Ape and the Sushi Master* (2001), primatologist Frans de Waal recounts a personal anecdote about the time he and Jan Van Hoof were not taken seriously when they suggested a study of chimpanzee 'reconciliations' in the 1970s (deWaal, 2001: 52). Jane Goodall's well-known refusal to number the chimpanzees she studied, preferring instead to name them, was a practice which ran counter to an objective masculinised orthodoxy of the time. In *Minding Animals* (2002) Mark Bekoff comments on the continuing taboos associated with treating animals in laboratories as anything other than objects (Bekoff, 2002: 49), and Pamela Asquith has written on anthropomorphism and the diverging 'conceptions of the human/animal relationship' in Japanese and western primate studies that, she notes, 'resulted in more than two decades' lag behind the Japanese in certain theoretical developments in primatology' (Asquith, 1997: 29). Anthropomorphism in these cases is a matter of both language and practice. These few examples also alert us to the various entangled power relations that involve anthropomorphism and which include but are certainly not limited to individual credibility and the professionalisation of scientific disciplines, gender essentialism, the reverence paid to ideas of rationality and objectivity over subjectivity, and the legacies of maintaining culturally specific constructions of human/animal difference. The justification or condemnation of anthropomorphism and disambiguation of its definitional vagaries delimits what can and cannot be said about animals and what can and cannot be done to them, as well as who is authorised to 'say' and 'do'.

Art and science

The contest of interpretation between the popular and the scientific is sidelined in many discussions about anthropomorphism where the spheres of arts and sciences are treated as exclusive. The problem with this approach, however, is that the public communication of science renders such exclusivity problematic. Those who have latterly come to endorse some forms of anthropomorphism language do so because of 'the lack of available alternative ways to describe animals and their behaviour, and the necessity of communicating in simple terms to lay audiences (including television viewers)' (Sealey and Oakley, 2013: 401). As the preceding discussion about discursive legitimacy suggests, a tendency to try to separate anthropomorphism in the realms of art and science resolves into traditional subjective/objective binaries. Stewart Guthrie's comprehensive account of anthropomorphism as a perceptual process expresses the distinction in this way: 'While scientists try to suppress it, creative writers and visual artists develop and use it' (Guthrie, 1993: 122). Anthropomorphism in the arts, he suggests, is 'intentional and contrived' (ibid.) while anthropomorphism in philosophy and science is subject to 'scepticism and scrutiny' (1993: 152). In his typology of anthropomorphism, John Andrew Fisher (1995) uses a branching diagram to illustrate the separation between 'imaginative anthropomorphism' and 'interpretive anthropomorphism', the latter of which is then subdivided into categorical and situational types. 'Imaginative anthropomorphism' according to Fisher is 'the productive activity of representing imaginary or fictional animals as similar to us'. He then goes on to give some examples: 'the animal characters in animations, books, drawings, movies and oral tales', which are bracketed away from the types of anthropomorphism that involve 'philosophers and scientists concerned about scientific explanations of animals' (Fisher, 1995: 6).

Frans de Waal also makes a distinction between science and the popular but argues that 'anthropomorphism [is not] necessarily unscientific, unless it takes one of the unscientific forms that popular culture bombards us with' (de Waal, 2001: 71). De Waal's typology distinguishes between naïve anthropomorphism – that he groups with paedomorphism (which endows animals with infantile features) and satirical anthropomorphism (which stereotypes animals for the purposes of mocking humans) – and 'animalcentric' anthropomorphism. Much popular culture, according to de Waal, exemplifies naïve anthropomorphism, which is 'profoundly anthropocentric' because it has little to do with what we know about the animals themselves and therefore 'attributes human feelings and thoughts to animals based on insufficient information or wishful thinking' (de Waal, 2001: 72–73). 'Animalcentric anthropomorphism', he argues, 'must be sharply distinguished from anthropocentric anthropomorphism. The first takes the animal's perspective, the second takes ours' (2001: 77). In the illustration of his classification of anthropomorphism, de Waal adds a further category, that of 'anthropodenial', into the mix. This, he proposes, is the opposite of anthropomorphism. It is, de Waal argues, the 'a priori rejection of shared characteristics between humans and animals when in fact they may exist' (2001: 69). For de Waal, it is anthropodenial that stresses distinct difference between humans and other animals while anthropomorphism can accommodate similitude, or as Mark Bekoff describes it, differences along a continuum (Bekoff, 2002: 48).

Randall Lockwood categorises anthropomorphism into five types: allegorical anthropomorphism, personification, superficial anthropomorphism, explanatory anthropomorphism, and applied anthropomorphism, which has some utility and benefit for science (Lockwood, 1989: 41–49). Allegorical anthropomorphism is most closely aligned with popular cultural representations of animals and includes 'fables, Disney films and literary or political allegories' which, as they do not 'presume to portray biological reality', are judged by Lockwood as 'essentially harmless' (1989: 45–46). Bekoff argues in favour of a particular type of anthropomorphism, one that he refers to as 'biocentric' and which, similar to de Waal's animalcentric type, can take the animal's point of view, thereby allowing 'other animals' behaviour and emotions to be accessible to us' (Bekoff, 2002: 48). Gordon M. Burghardt also finds much merit in a particular type of anthropomorphism that he identifies as 'critical',

> in which our statements about animal joy and suffering, hunger and stress, images and friendships, are based on careful knowledge of the species and the individual, careful observations, behavioral and neuroscience research, our own empathy and intuition, and constantly refined publicly verifiable predications.
>
> (Burghardt, 1997: 268)

For Burghardt, the aim is to pose 'questions anthropomorphically, but [think] about them critically' (1997: 269).

These typologies reflect a range of concerns to do with language, methodology, epistemology, human exceptionalism and ethics. That many of the debates about anthropomorphism and corresponding typologies emerge during the latter part of the twentieth century is of interest in terms of periodisation, signalling a shift away, in some disciplinary areas, from conceptualising anthropomorphism as a single 'innate' practice, instead fragmenting it into different types and forms. This plurality of anthropomorphisms has marked a challenge to hyper-differentiation between humans and other animals, opposing to some degree the anthropocentric dualism that underpinned anthropomorphism's pejorative status in scientific disciplines. But it would be overstating the influence of these debates on anthropomorphism to suggest that they constitute a 'turn'; anthropomorphism remains controversial, and what de Waal refers to as anthropodenial still holds plenty of ground. Despite a renewed focus on the complex mental lives of animals since the late twentieth century, consciousness in animals other than humans remains contentious. Even though 'strong claims' about animal mentality have been made, as Burghardt (2009) observes, this raises 'again the problem of anthropomorphic interpretations of behavior that, when uncritically deployed, open up opportunities for critics to tar all attempts to study the evolution of mental experience' (Burghardt, 2009).

There is no doubt that it has been incredibly important for ethologists who have endorsed anthropomorphism and its methodological utility to carve out some distinctions between the scientific and popular realms. This is very much a continuation of earlier epistemological struggles to establish the professionalism of the sciences and identify the particularities of knowledge formations. Where the discourse of serious science has located anthropomorphism as the stuff of subjective popular entertainment, ethology has necessarily found a way of acknowledging its methodological relevance while maintaining a distance from the anthropomorphism of commercial entertainment. At the same time, typologies of anthropomorphism reinforce problematic fault lines of difference between science and popular culture. Typologies suggest that we can think of anthropomorphism in popular culture as easily grouped into one category, as 'harmless', as subject to critique on the basis that anthropomorphism is inaccurate, not 'objective' or perhaps even left unhampered by the criticisms of science.

Although it might claim that there is more at stake, science has not monopolised the debates over anthropomorphism. On the face of it, popular culture might be considered an uncontentious sphere for anthropomorphic practices. In the realm of the 'popular', talking, thinking, feeling animals – or at least their representations – can be enjoyed and even applauded. However, the perceived safe confinement of anthropomorphism to this sphere has marshalled meanings that contribute to its derogatory status in other fields. Randall Lockwood, who argues for the benefits of anthropomorphism for science, writes about his entry into the scientific study of animal behaviour and his 'distinct impression that it was time to put away my teddy bears and memories of Disney films and acquire the cold, hard stare of the "objective" scientist' (Lockwood, 1989: 41). The explicit distancing from Disney becomes something of a theme when anthropomorphism is the topic at hand. For instance, 'Bambification' has been used to refer to overly sentimentalised forms of anthropomorphism (deWaal, 2001), and Disney often figures as the go-to cultural reference point for anthropomorphism, the suffix stressing the unwanted influence of 'the popular' on human understanding of other animals. The term 'Disneyfication' is commonly used to describe 'the assignment of some human characteristics and cultural stereotypes' to animals (Bekoff, 2009: 173). Disneyfication and anthropomorphism have become entwined and where they are considered primarily confined to popular culture, concerns have been raised about the impacts of inaccurate nonhuman animal representations on children, the public understanding of other animals, or even future generations of scientists (Kennedy, 1992). While it might be acknowledged that Disneyfication is not restricted to Disney products, the practice is thought to be exemplified in the company's output. Disneyfication thus has the dubious claim to being a highly potent and popular anthropomorphic practice that has little to recommend it, and is broadly regarded as tricky from scientific, animal rights and welfare perspectives.

Disney as a shorthand for anthropomorphism attests to both the cultural dominance of the company in relation to the production of animal-focused media, and a perception that there remains a consistency of style and aesthetics in their output. There is a danger in reducing anthropomorphism to the notion of Disneyfication which cannot easily take account of the multiple contact points between science and the popular in, for instance, news, documentary, wildlife films and other so-called factual genres. It is here where 'science' as a proxy for rational objectivity or 'common sense' can be invoked to undermine public empathy for other animals. Moreover, Disneyfication cannot account for the multiple ways in which anthropomorphism is deployed within popular culture. For this reason, in the case of popular culture, the scientific construction of anthropomorphism should be regarded as a regulating force and not as a measurement of 'validity' for popular anthropomorphism. To acknowledge the diversity not only of anthropomorphism but also the complexity of its production and reception in popular culture and, crucially, the different ways in which it is evoked in relation to individual nonhuman animals, anthropomorphism is, I suggest, better regarded as situational, contextual, differentiated and entangled. I have so far provided one set of conditions by which anthropomorphism in popular culture can be situationally constructed in a pejorative sense, and I apply this in case study examples in later chapters. In the next part of this chapter I deal with the other entangled sites of commodified anthropomorphism, affective labour and empathy.

From science to popular culture

What would it mean for a nonhuman animal to be represented on their 'own terms', who would make the decision about how another being's experience should be represented and, to apply the logic of the commercial media marketplace, would there be an audience for such narratives? The notion of truthful representations is often bound up with issues of genre where some forms of nonhuman animal representation are considered to have greater authenticity than others. The distinction between 'nature films' and fictional family features, for example, gives context to the ways in which we think about and consume anthropomorphic depictions.

The relationship between genre and expectations about 'truthful' representations of animal behaviour is made apparent in the case of the live action Disney True-Life Adventures series of films. As Gregg Mitman observes, 'If critics objected to the anthropomorphic conventions found in the True-Life adventures, it was only because Disney claimed the pictures to be real' (Mitman, 1999: 120). Yet, the critical gatekeepers in this case were film critics and not naturalists. Mitman discusses how the relationship between 'artifice and authenticity' in the case of nature films is far from simple and notes that conservation organisations endorsed the mid-twentieth-century Disney nature films. A similar situation arose in the 2000s when Disney created Disneynature, a semi-autonomous film unit dedicated to wildlife filmmaking. Aligning each film with a conservation charity, the highly sentimental and anthropomorphic Disneynature wildlife films received endorsements from high-profile organisations, which, in turn, benefitted from the film's public donation schemes. The point here is that while genre may cue certain expectations it does not guarantee 'truthful' or non-anthropomorphic representations of nonhuman animal lives and experiences. David Whitley makes an important point in this regard when he analyses the documentary *March of the Penguins* and the animated family entertainment film *Happy Feet*. 'Oddly', he observes, 'it is the latter film that bears the weight of cultural anxieties concerned with lost or degraded environments much more explicitly. The lightweight fantasy appears to be carrying a heavier load' (Whitley, 2014: 146). What Whitley's analysis makes clear is that a non-realist aesthetic does not militate against the communication of real environmental concerns from a nonhuman animal perspective.

Although the trend for anthropomorphism in wildlife documentary films has received scholarly attention, the go-to referents 'Disney' and 'Bambi' suggest that it is animated films which remain synonymous with anthropomorphic practices. There is good reason for this in the sense that anthropomorphism is well established and discussed in the animation industry as an interpretive artistic practice. Strangely perhaps, a focus on anthropomorphism has not given rise to much critical engagement in animation studies with the 'question of the animal'. Paul Wells makes this point when he notes that despite 'the ubiquity of the animal in animation since its early beginnings, it has not been a consistent preoccupation for analysis' (Wells, 2009: 2). Jonathan Burt makes a similar observation, noting that the general film books he consulted while writing *Animals in Film* might have carried index entries for 'animation' and 'anthropomorphism', yet it was rare to 'find specific index entries for animals' (Burt, 2002: 17). Wells reasons that 'there is an almost taken-for-granted sense about animals in animation'. He goes on to propose that 'the animal is an essential component of the language of animation, but one so naturalised that the anthropomorphic agency of creatures [. . .] has not been particularly interrogated' (2009: 2).

Soviet director and film theorist Sergei Eisenstein intellectualised in celebratory tones the relationship between animism and anthropomorphism in animation – a union between primitive archaic ritual and art practices. For Eisenstein, the tendency for ' "flight" into animal skins and the anthropomorphic qualities of animals' occurred across epochs, reliant on differentiation, or rather the collapse of distinctions, between the objective and the subjective, between our 'knowing' and our 'sensing'. In the case of Disney, he wrote:

> We *know* that these are drawings, and not living beings. We *know* that this is a projection of drawings onto a screen. We *know* that they are 'miracles' and tricks of technology and that such beings don't actually exist in the world.
>
> (Eisenstein, 1986: 55)

And, he exclaims, at the same time, 'We *sense* them as living, we *sense* them as active, acting; we *sense* them as existing and we assume that they are even sentient!' (Eisenstein, 1986: 55).

The substitution of animals for humans in a Disney animated cartoon was, Eisenstein argued, 'practically a direct manifestation of the method of animism' and 'one of the deepest features of the early human psyche' (1986: 43). For Eisenstein, the subjective realm of sensuous thinking that characterised animism put humans into a different relationship with nature, one that was open to the interconnectedness of all things, 'all elements and kingdoms of nature' (1986: 52). Objective examination ruptured this interconnection by emphasising difference; animism, 'the "blissful" condition of [. . .] sensory experience' by way of contrast, sought similitude by 'analogy with the personal' (ibid.). In Eisenstein's theorisation of popular animation, animism and, by extension, anthropomorphism is granted a certain legitimacy that acted as a counter to science's objections about the innate anthropomorphic bias. Animistic belief was placed in contrast with science and 'the objective examination of the surrounding world' with 'its sequences and stages' (ibid.). Wells argues that Eisenstein 'recognized that the animal story has played a significant role in liberating humankind from its inhibitions and limitations in specific historically determined moments' (Wells, 2009: 63). Taking this argument further, Wells proposes that in relation to the development of animated film, 'it is clear that the animal is the embodiment not merely of motion itself as the core characteristic of animation practice, but the carrier of notions of movement as the signifier of social change' (Wells, 2009: 65). For Wells, animated animals can play with radical ideas about human social identities in ways denied to other media at certain moments in history. Christopher Holliday similarly argues that 'anthropomorphism has thus remained inextricably bound to animation's potential as a subversive, immanently creative art' (Holliday, 2016: 25).

Eisenstein referred to the protean qualities of animation as 'plasmaticness' and revelled in the medium's ability to turn an octopus into an elephant and 'dynamically assume any form' (Eisenstein, 1986: 21). What does this mean, though, for the figure of the animal? Is she simply a mutable form, easily erased? According to Holliday, 'the animated anthropomorph acts as an outlet for this particular brand of aesthetic freedom, every frame a confirmation of its achievements and possibilities' (Holliday, 2016: 25). Wells argues that 'the animal' is not erased by the process of becoming a human metaphor, and he identifies examples where 'the animator's empathetic engagement with [. . .] felt experiences in the animal [. . .] offer the potential to depict the most authentic idea of the pure animal' (Wells,

2009: 117). For Wells, animation can reveal human connections with animality where anthropomorphism can function as an 'interrogative tool' that engages with the potential relationships between human and nonhuman animals (Wells, 2009: 96). Eisenstein's and Wells' commentaries both point towards a more open and relational form of anthropomorphism in animation, one that does not reduce animals to simple human caricatures but instead reconnects humans in a sensuous way to a subversive, animistic way of viewing the world.

Animism – attributing life to something that is not alive – can result in both anthropomorphic and zoomorphic interpretations, a point that has resulted in some academic speculation over whether a popular animated character such as Mickey Mouse is a mouse with human characteristics or a human with mouse characteristics (Lawrence, 1989; Guthrie, 1993; Avis, 2014). Where do anthropomorphism and zoomorphism start and end, and why do these representations leave us with so much definitional discomfort? For Berger (1980), Mickey Mouse is anthropomorphic, with all of the pejorative weight that the term brings with it. Stewart Guthrie remains ambivalent: 'Looking at Mickey Mouse, we do not know whether we see a mouse in man's clothing or a man in a mouse's body. We probably see both' (Guthrie, 1993: 134). Elizabeth A. Lawrence disagrees, and although she argues that while the character of Mickey is the 'most highly anthropomorphized of all fanciful animal characters', she contends that 'virtually no one perceives this familiar creature as a mouse!' (Lawrence, 1989: 58). Avis, on the other hand, regards Mickey as a blend of mouse and human and argues that there is an evolutionary advantage to being able to see 'cross domain perceptual realities' (Avis, 2014: 56). In this regard, for Avis, category blending is an example of the cognitive fluidity that enables abstract thinking and creativity.

Steve Baker divides zoomorphism into therianthropism and therianmorphism and offers some important definitional clarification, pointing out that 'a theriomorphic image would be one in which someone or something [. . .] was presented as "having the form of a beast". Therianthropic images', Baker notes, combine 'the form of a beast with that of a man' (Baker, 2001: 108), and these seldom used terms more accurately describe 'the casting of humans into animal form' than the 'frequently misused notion of anthropomorphism' (ibid.). According to Baker, Disney exploits the anthropomorphic and therianmorphic and 'appears to care very little about the distinction between these two classes of representation' (Baker, 2001: 226). Baker reads zoomorphism across the textual and the extra-textual, arguing that while Disney animal characters are anthropomorphic, the humans dressed up as those characters – in Disney theme parks for example – are therianthropic. Although the distinction between the two might be muddied by the constructed experience of the theme park, they take on different meanings when appropriated for other causes. In the case of the iconic 'Mickey ears', which for Baker are a 'willing engagement with the therianthropic', when used against their 'stereotypical meanings' by, for example, animal rights protestors, the Disney imagery shows 'surprising potential for manipulating and undermining stereotypical conceptions of the animal' (Baker, 2001: 228–229). For Baker, then, the Mickey ears are therianthropic but their meaning is unfixed and dependent on context. In other words, powerful anthropomorphic imagery can be subverted through appropriation such that the capitalist mouse – or in Baker's terms, certain therianthropic assimilations of Mickey mouse-ness – can also be symbols of vulnerable animality.

Sean Cubitt (2005) argues that animation makes possible trans-species identification through a combination of anthropomorphism and zoomorphism. Similar to Wells, Cubitt also argues that to become part of society, children must learn to separate animals from humans. Animals then are something which develop a child's understanding of difference and solidify their sense of human identity. He writes, 'This is the process of internalising the chiasmus of Agamben's anthropogenic machine, the systems of self-awareness which will, constantly revising and rewriting the distinction, generate the humanness of the human being out of the difference from animality' (Cubitt, 2005: 30). Cubitt argues that even the most anthropocentric of anthropomorphic depictions retains some memory of animality. 'Identification', Cubitt argues, 'is thus extended beyond mere empathy into a realm where the relationship is transgressed not by the projection of human qualities onto anthropomorphised animals, but by the introjection of animal properties into the spectator: the process of zoomorphism' (Cubitt, 2005: 32). Cubitt, like Wells, considers the relationship of the animator to 'the animal' in the process and argues that the animator must feel what it is to be the animal to successfully draw the animal. Therefore, there is a double relationship wherein the animator must open themselves up to zoomorphic introjection to enable their production of anthropomorphic projection (Cubitt: 2005: 32–36).

As Wells and Cubitt suggest, it is important that the processes of production, the humans (and, I argue, the nonhumans) who are involved in those processes and the means by which the animal is mediated are taken into account to ensure that we don't think only about the textual or the representational as somehow divorced from the economic logics and practices of the cultural and media industries. Moreover, even if we take anthropomorphism as generally reflecting human concerns when it occurs in commercial media *and* we acknowledge that those human concerns do include the perspectives, welfare, rights, experiences or lives of nonhuman animals in some form, then it stands to reason that anthropomorphism does more than merely simplify or erase the 'real' nonhuman animal from the narrative. Some would certainly disagree with this. In his critique of anthropomorphism, those who are responsible for humanised representations of nonhuman animals are notoriously referred to by Ralph H. Lutts as 'nature fakers'. Lutts argues that animal welfare advocates, nature writers and nature fakers pioneered in the 1980s a new approach to encourage public concern for wild nonhuman animals. This included offering 'sympathetic accounts about the lives of individual animals' which presented 'their animal heroes as furry or feathery little people' (Lutts, 1990: 202). The issue for Lutts is that this results in a type of logical fallacy: the public only care about animals they understand as 'human'; the 'authors' have to present the animals as 'human' to garner public support; the animals were 'caught in plots that human readers could relate to. But their protagonists were not human' (ibid.). For some filmmakers, however, anthropomorphism is the process by which complex nonhuman animal lives can be translated for a general audience. As filmmaker Sarita Siegel writes in an essay about her documentary *The Disenchanted Forest*, 'Anthropomorphic metaphors, anecdotes and analogies are extremely useful when combining image and narration as a means of portraying complex orangutan "personalities", who might otherwise be seen as unremarkable when viewed by an audience untrained in observing such complex creatures' (Siegel, 2005: 197). Siegel's approach contradicts Lutts' position and is representative of a move towards more overt anthropomorphism in nature and wildlife programmes in the 1990s and 2000s. Cynthia Chris (2006) contextualises the late twentieth-century shifts in wildlife programme making and their use of anthropomorphic cues. Placing the genre conventions of the wildlife film in relation to other literary traditions, Chris identifies how film and programme makers tried to address audiences in a changing media landscape, characterised by the fragmentation of audiences, the growth of niche channels and expansion into global markets. Derek Bousé, in his account of the history wildlife films, identifies that anthropomorphism in mass media representations of nature can be traced back to the nineteenth-century nature writing, a precursor to the wildlife films of the twentieth century. Bousé crucially also makes a link between human concern for other species and anthropomorphism when he notes that nineteenth-century animal welfare was 'closely tied, if not dependent on, some degree of anthropomorphism' (Bousé, 2000: 99). Popular anthropomorphism thus has a history of engaging audiences in concern for other species and media makers understand very well that the affective pulls of anthropomorphic animals can have benefits beyond the economic.

Commercial media

In the study of media and popular culture, Disneyfication has a different definition to that discussed earlier and refers to processes of homogenisation, 'sanitization and Americanization' (Wasko, 1998) – in other words, with 'how Disney bowdlerizes history, culture, myth, and literature into streamlined, charming, and pre-digested forms devoid of content that might be considered noxious or offensive to the consumer of Disney products and images' (Kalin in Sandlin and Garlen eds., 2016: 200). In eco-cinema studies, as Brereton points out, there exists a whole strand of academic criticism concerned with 'Disneyfication' and its correlative 'reductive anthropomorphic simplification of nature' (Brereton, 2014: 183). There is an understandable apprehension about Disney and its mediation of nature if, as Brereton points out, one is inclined towards ideological analysis and arguments that conceive of the audience for popular culture as an undiversified mass that absorbs uncritically 'the commercially potent, yet synthetic and sentimental, nature of the Disney [. . .] project' (ibid.). Aligned with this perspective is a general critique of the culture industries which positions media conglomerates such as Disney as capitalistic enterprises that manipulate consumers and influence cultural values and norms. The long-standing history of such concerns within studies of popular culture is perhaps most forcefully articulated by Theodor W. Adorno and Max Horkheimer in their vehement critique 'The Culture Industry: Enlight-enment as Mass Deception', where they condemn the standardisation of mass culture and its ideological domination, arguing that, even where there is product differentiation, this is to ensure consumers are classified, organised and labelled. In other words, they contend, 'something is provided for all so that none may escape' (Adorno and Horkheimer, 1997: 123). That 'something' – mass culture – in Adorno and Horkheimer's view purposefully supresses the critical capacities of consumers.

Commercial media is by its very definition a capitalist enterprise. Capitalism as an economic system does not have a moral agenda and nonhuman animals have historically fared very poorly under its organisation of social and cultural life. Indeed, from a political economy viewpoint, nonhuman animals have been exploited mercilessly to further capitalism within most if not all sectors, and com-mercial media has taken a key role in normalising that exploitation. Anthropo-morphism is one of the tactics used in marketing, cinema, television, social media and elsewhere, making it on one level a disconcerting assistant to the cruelty and oppression of nonhuman animals. It is easy to be critical of anthropomor-phism conceived of in this way. Taking this position on the culture industries and consumers, Disney's anthropomorphism reinforces human social norms, val-ues and morals in a form of profitable escapist entertainment which leaves little, if any, space for resistant readings. The 'mass' audience, in its exposure to the dominance of Disney, is thereby indoctrinated into understanding other animals through a specific type of anthropomorphic lens, one that sees them as little more than 'humans in fur coats'. Audiences are infantalised and nature is constructed as pristine, innocent and unchanging. Disneyfication thus becomes a signifier for all that is wrong with the worst excesses of anthropomorphism in popular culture.

Such positions have not gone unchallenged. A study of the effects of anthropo-morphism on children's knowledge about real nonhuman animals concluded that 'anthropomorphic language and pictures in storybooks did not interfere with fac-tual learning about real animals' (Geerdts et al., 2015). In his defence of Disney-fication, Paul Wells re-reads the infantalisation of the audience, through Freud, as a way for adults to reconnect with nature. For Wells, there is a primal connection between children and animals that is lost as they move into adulthood. Animation generally and Disney cartoons specifically are an opportunity to recall what Wells refers to as 'primal bonds and lost knowledge' (Wells, 2009: 78). He argues,

> Modern estrangement in adults, then, might be understood as not merely alienation from nature, but from the ease of correspondence and understand-ing of animals that comes with the negation of hierarchy and the empathetic needs that forestall superficial questions of difference.
>
> (Wells, 2009: 78)

Pat Brereton is also critical of what he refers to as the 'paedocratizing' of the Disney audience (Brereton, 2014: 185). Here, Brereton refers to John Hartley's critique of the television industry and its motivation to *paedocratize* audiences: a set of practices that involve imagining the audience to have 'childlike qualities and attributes' and therefore 'addresses its viewers as children' to sustain profitability and minimise risks (Hartley, 1992: 108). Brereton argues that there isn't a singular type of Disney anthropomorphism and that Pixar, a subsidiary of the Walt Disney Studios (the film studio division of The Walt Disney Company), has been able to subvert the 'branded studio model of Disney' to the extent that 'while this so-called "Disneyfication of nature" might continue to display a reductive anthropomorphic simplification of nature, it becomes at least potentially progressive by generating smart ecological metaphors for audiences to engage with' (Brereton, 2014, 181–183). Brereton also contends that commercial filmmakers have expanded their repertoire, 'broadening and deepening their narrative address and enticing mass audiences to extend their engagement with a more radical form of ecological engagement' (2014: 185). If it constantly focuses on the twin tropes of childhood innocence and nature, ideological criticism can, Brereton proposes, ignore the 'eco-utopic significance and emotional affect' of eco-narratives (186). In this sense, Brereton rightly points out that Disney and the process of Disneyfication are low-hanging fruit when it comes to academic criticism. It is perhaps easy to agree with Elizabeth A. Lawrence's observation that Mickey Mouse tells us almost nothing about mice but 'a great deal about ourselves' (Lawrence, 1989: 75). As iconic as they might be, however, we should not reduce anthropomorphism, Disney's output, or for that matter the medium of animation, to Mickey or Bambi. If we take Disney as a proxy for popular culture's overstatement of similitude between human and nonhuman animals, we risk ignoring the situational and contextual diversity of popular anthropomorphism and dismissing its affective dimensions as 'merely' over-sentimentalised.

Disney, like many other companies in the fields of entertainment as well as in other industry sectors, invests much in anthropomorphism to engage human consumers with products, narratives and other forms of animal commodification. When I refer to sites of commodified anthropomorphism, I am thinking specifically about sites where anthropomorphism is employed to distance humans from the material reality of other animals' lives. In this sense, anthropomorphism is deployed as a mechanism of commodification to construct an illusory sense of animal agency. It is also the case that these sites construct animals as affectively engaging and that the affective labour of animals is commodified in ways that serve wholly anthropocentric interests. At the same time, these sites of anthropomorphism are contextual, their meanings are not fixed. A site of commodified anthropomorphism and the affective labour of animals can also mobilise empathetic connections for audiences and consumers, if not directly, then indirectly if we think about these sites as entangled within popular culture. Jonathan Burt suggests that 'certain kinds of animal imagery, magnified and intensified precisely by the artifice of film, are responded to more emotionally and are therefore less mediated by the judgements that we might normally apply to other types of imagery' (Burt, 2002: 10). Family films, Burt notes, come under particular critical scrutiny 'on the grounds that their sentimentalism and anthropomorphism create a comfortable complacency in attitudes to animals' (Burt, 2002: 187). Much like 'Disney', family films are a target for criticism when it comes to anthropomorphism precisely because they have overstated affective qualities. However, audiences respond to nonhuman animals in commercial media sometimes in unexpected ways, and there are plenty of examples of public outrage over the treatment of certain animals depicted in films (see Burt, 2002; Loy, 2016 and this book). This suggests that where commodified anthropomorphism is culturally encouraged, vulnerabilities in the capitalistic exploitation of animals are also exposed. By this I mean that where consumers are encouraged to emotionally invest in commodified anthropomorphism to sell an 'animal product', whether that be a 'happy cow' to sell milk or a commercial film narrative about a loyal dog, commodified anthropomorphism can be vulnerable to empathetic anthropomorphism, particularly where the same animals are shown to be exploited or cruelly treated. In terms of the economics of such dynamics, the reputational capital of companies that derive profit from the affective labour of animals is at risk if those businesses are shown to exploit the animals that they ask consumers to emotionally invest in.

Anthropomorphic affect

The issue of audience address and engagement with nature films and television programmes is one that has prompted critical consideration of the use of anthropomorphism in communicating environmental, conservation, animal rights and welfare messages. Bart H. Welling, for example, argues that 'wildlife films can [. . .] provide viewers with heavily mediated but potentially transformative modes of access to the emotional lives of our non-human kin' (Welling, 2014: 82). Robin L. Murray and Joseph K. Heumann also argue that affective appeals and anthropomorphic depictions might be more successful in engaging an audiences and effecting change compared to films which eschew such approaches in favour of a 'wise-use' eco-narrative. They argue that the impact of films can be increased by humanising nonhuman individuals, 'demonstrating that they, like humans, have rights' or providing greater 'emotional connections' between the audience and the subject (Murray and Heumann, 2014: 136–137). Although not concerned with film as such, John Law and Marianne Elizabeth Lien discuss the impossibility of finding a non-anthropomorphic language. The issue they propose is 'not so much anthropomorphism itself – this cannot be avoided – but rather how it might best be *done*' (Law and Lien, 2017: 42). Adrian Ivakhiv also writes about the inevitability of anthropomorphism, in this case in relation to films, because cinema will always tell stories from a human perspective. He reasons, 'because film shows us human or human-like subjects, beings we understand to be thrown into a world of circumstance and possibility like us, it is *anthropomorphic*' (Ivakhiv, 2013: 9). Some of this work makes the case for rethinking anthropomorphism in more productive ways – as relational, functional, engaging, affective and effective.

The criticisms of anthropomorphism in popular culture and the dismissiveness with which it has been treated, grouped under the pejorative umbrella term 'Disneyfication', have ignored or denigrated its affective dimensions. This often means falling into the trap of thinking about emotional engagement with popular narratives as manipulative, lacking criticality and at odds with the notion of a rational adult audience. Popular culture may indeed produce a standardisation of experience, one which is reductive in the sense that, to be commodified for appeal to the widest audience, it utilises familiarity and repetition. This should not be taken to suggest, however, that audiences are passive or that their critical capacities are suppressed. Popular culture places other animals within a human framework, one of familiarity that reproduces the nonhuman animal as humanised and easily commodified by capitalistic enterprises. But if we consider the affective dimensions of popular culture, the effects of this humanisation can be problematic for an economy built on the exploitation of animals. Popular culture produces surpluses of similitude, the affective dimension of which can reconnect humans at an empathetic level to other animals as the imagined objects of capitalistic culture which, under certain conditions, are depicted as subjective beings. This process can grant nonhuman animals agency beyond the confines of the text where they might motivate empathetic concerns, enhance relational understanding and shape transformative connections. This surplus of similitude can of course be commodified, for example in the merchandising of animal characters and the sale of exotic species through the pet trade, but it can also engage humans, not only as consumers, but as beings with experiences that are coextensive with those of other animals. The conditions that reduce animals to commodified humanised objects of popular culture can also enable imaginative connections with animals that have unexpected outcomes.

Casting all anthropomorphism into the category of problematic sentiment belittles affect but also fails to address popular narratives as historically situated and addressing the public concerns, debates or anxieties of their time. Popular narratives and other forms of human-to-human communication will inevitably be anthropomorphic. Cinema and television particularly imbue characters with agency and intentionality. In other words, anthropomorphism is inevitable within popular culture. We can of course be critical of anthropomorphism and argue that it erases difference, simplifies the complexity of nonhuman animal experience and imposes a human framework on the more-than-human world – a last gasp of anthropocentric force as we face the consequences of human arrogance writ large as the Anthropocene. However, I do not think that an all-out rejection of anthropomorphism is needed or helpful. Anthropomorphism is enmeshed in power relations that have sought to denigrate affect through the feminisation of emotion, to posit infantalisation in opposition to the enlightenment project of rational adulthood and to position animism as a form of 'primitive' knowledge that opposes the interests of a white western masculinised science. Anthropomorphism cannot be disentangled from the historical binaries that have reinforced a general aversion to its presence. It is vital, however, that we are attentive to those power relations and do not overlook the affective potential of anthropomorphism.

References

Adorno, T. W. and Horkheimer, M. (1997) *Dialectic of Enlightenment*, Verso, London and New York.

Asquith, P. J. (1997) 'Why anthropomorphism is not metaphor: Crossing concepts and cultures in animal behavior studies' in Mitchell, R. W., Nicolas, S. T., and Lyn Miles, H. (eds) *Anthropomorphism, Anecdotes, and Animals*, SUNY Press, New York, pp. 22–36.

Avis, M. (2014) 'Cross domain perceptual realities and Mickey Mouse' in Brown, S. and Ponsonby-McCabe, S. (eds) *Brand Mascots and other Marketing Animals*, Routledge, New York and London, pp. 55–76.

Baker, S. (2000) *The Postmodern Animal*, Reaktion Books, London.

Baker, S. (2001) *Picturing the Beast: Animals, Identity and Representation*, University of Illinois Press, Chicago.

Barnett, L. (2012) 'A palaeontologist's view on ice age: Continental drift' *The Guardian* 23 July, online at www.theguardian.com/film/2012/jul/23/palaentologist-view-ice-age-continental

Beer, G. (1984) *Darwin's Plot: Evolutionary Narrative and Nineteenth Century Fiction*, Routledge, London.

Bekoff, M. (2002) *Minding Animals: Awareness, Emotions, and Heart*, Oxford University Press, Oxford.

Bekoff, M. (2007) *The Emotional Lives of Animals*, New World Library, California.

Bekoff, M. (ed) (2009) *Encyclopedia of Animal Rights and Animal Welfare*, Routledge, London and New York.

Berger, J. (1980) *About Looking*, Random House, New York.

Bousé, D. (2000) *Wildlife Films*, University of Pennsylvania Press, Philadelphia.

Brereton, P. (2014) 'Animated ecocinema and affect' in von Mossner, A. W. (ed) *Moving Environments: Affect, Emotion, Ecology and Film*, Wilfred Laurier University Press, Ontario.

Brereton, P. (2016) *Environmental Ethics and Film*, Earthscan, New York and London.

Burghardt, G. M. (1997) 'Amending Tinbergen: A fifth aim for ethology' in Mitchell, R. W., Thompson, N. S., and Miles, H. L. (eds) *Anthropomorphism, Anecdotes, and Animals*, SUNY Press, Albany, NY, pp. 254–276.

Burghardt, G. M. (2009) 'Ethics and animal consciousness: How rubber the ethical ruler?' in *Journal of Social Issues*, Vol. 65 (3), pp. 499–521.

Burt, J. (2002) *Animals in Film*, Reaktion Books, London.

Chris, C. (2006) *Watching Wildlife*, University of Minnesota Press, Minneapolis and London.

Connor, S. (2004) 'Ice age movie is realistic, says Britain's chief scientist' *Independent*, 13 May, online at www.independent.co.uk/news/science/ice-age-movie-is-realistic-says-britains-chief-scientist-563199.html

Crist, E. (1999) *Images of Animals: Anthropomorphism and Animal Mind*, Temple University Press, Philadelphia.

Cubitt, S. (2005) *EcoMedia*, Rodopi, Amsterdam and New York.

Daston, L. and Mitman, G. (eds) (2005) *Thinking with Animals: New Perspectives on Anthropomorphism*, Columbia University Press, New York.

Dawkins, M. S. (2012) *Why Animals Matter: Animal consciousness, Animal Welfare, and Human Well-Being*, Oxford University Press, Oxford.

de Waal, F. (2001) *The Ape and the Sushi Master: Cultural Reflections of a Primatologist*, Basic Books, New York.

Eisenstein, S. M. (1986) *Eisenstein on Disney* [trans Alan Upchurch], Seagull Books, Calcutta.

Fieser, J. (2001) *Moral Philosophy Through the Ages*, Mayfield Publishing, London and Toronto.

Fisher, J. A. (1995) 'The myth of anthropomorphism' in Bekoff, M. and Dale, J. (eds) *Readings in Animal Cognition*, MIT Press, Cambridge, MA and London, pp. 3–16.

Geerdts, M., Van de Walle, G.A., and Vanessa, L. (2015) 'Learning about real animals from anthropomorphic media' in *Imagination, Cognition and Personality*, Vol. 36 (1), pp. 5–26.

Greenberg, G. and Haraway, M. (1998) *Comparative Psychology*, Garland Publishing Inc, New York and London.

Gruen, L. (2015) *Entangled Empathy: An Alternative Ethic for our Relationships with Animals*, Lantern Books, New York.

Guthrie, S. (1993) *Faces in The Clouds: A New Theory of Religion*, Oxford University Press, Oxford.

Hartley, J. (1992) *Tele-Ology: Studies in Television*, Routledge, London and New York.

Holliday, C. (2016) 'Carl's moving castle: "Animated" houses and the renovation of play in *Up* (2009)' in Andrews, E., Hockenhull, S., and Fran P-K. (eds) *Spaces of the Cinematic Home: Behind the Screen Door*, Routledge, London and New York, pp. 19–31.

Ivakhiv, A. J. (2013) *Ecologies of the Moving Image: Cinema, Affect, Nature*, Wilfred Laurier University Press, Ontario.

Kalin, N. M. (2016) 'Disneyfied/ized participation in the art museum' in Sandlin, J. A. and Garlen, J. C. (eds) *Disney, Culture, and Curriculum*, Routledge, New York and London, pp. 193–207.

Kennedy, J. S. (1992) *The New Anthropomorphism*, Cambridge University Press, Cambridge.

Law, J. and Lien, M. E. (2017) 'The Practices of Fishy Sentience' in Asdal, K. D. and Steve, H. (eds) *Humans, Animals and Biopolitics: The More-than-Human Condition*, Routledge, London and New York, pp. 30–47.

Lawrence, E. A. (1989) 'Neotony in American perceptions of animals' in Hoage, R. J. (ed) *Perceptions of Animals in American Culture*, Smithsonian Institution Press, Washington, DC and London, pp. 57–76.

Lockwood, R. (1989) 'Anthropomorphism is not a four-letter word' in Hoage, R. J. (ed) *Perceptions of Animals in American Culture*, Smithsonian Institution Press, Washington, DC and London, pp. 41–56.

Loy, L. (2016) 'Media activism and animal advocacy: What's film got to do with it?' in Almiron, N., Cole, M., and Carrie, P. F. (eds) *Critical Animal and Media Studies: Communication for Nonhuman Animal Advocacy*, Routledge, New York and London.

Lutts, R. H. (1990) *The Nature Fakers: Wildlife, Science & Sentiment*, University Press of Virginia, Charlottesville and London.

Mead, G. H. (1907) 'Concerning animal perception' in Mead, G. H. and Reck, A. (ed) *Selected Writings*, Bobbs-Merrill, Indianapolis, pp. 73–81.

Midgley, M. (1998) *Animals and Why They Matter*, University of Georgia Press, Athens.

Mitman, G. (1999) *Reel Nature: America's Romance with Wildlife on Film*, University of Washington Press, Seattle and London.

Morgan, C. L. (1903) *An Introduction to Comparative Psychology* (Revised Edition), Walter Scott Publishing, London.

Murray, R. L. and Heumann, J. K. (2014) *Film and Everyday Eco-disasters*, University of Nebraska Press, Lincoln and London.

Pavlov, I. P. (1963) *Lectures on Conditioned Reflexes Vol II: Conditioned Reflexes and Psychiatry*, Lawrence and Wishart, London.

Pfungst, O. [1907] (1998) *Clever Hans (The Horse of Mr. von Osten)*, Theommes Continuum, Bristol.

Plumwood, V. (2002) *Environmental Culture: The Ecological Crisis of Reason*, Routledge, New York.

Ritvo, H. (1987) *The Animal Estate: The English and Other Creatures in the Victorian Age*, Harvard University Press, Cambridge, MA.

Ryder, R. D. (1989) *Animal Revolution: Changing Attitudes Towards Speciesism*, Blackwell, London.

Sealey, A. and Oakley, L. (2013) 'Anthropomorphic grammar? Some linguistic patterns in the wildlife documentary series *Life*' in *Talk & Text*, Vol. 33 (3), pp. 399–420.

Seward, A.C. (1909) *Darwin and Modern Science*, Cambridge University Press, Cambridge.

Siegel, S. (2005) 'Reflections on anthropomorphism in *The Disenchanted Forest*' in Daston, L. and Mitman, G. (eds) *Thinking with Animals: New Perspectives on Anthropomorphism*, Columbia University Press, New York, pp. 175–195.

Thorndike, E. (1911) *Animal Intelligence*, Palgrave Macmillan, New York.

Turner, J. (1980) *Reckoning with the Beast: Animals, Pain, and Humanity in the Victorian Mind*, John Hopkins University Press, Baltimore and London.

Wasko, J. (1998) *Understanding Disney: The Manufacture of Fantasy*, Wiley, New York.

Watson, J.B. (1913) 'Psychology as the behaviorist views it' in *Psychological Review*, Vol. 20 (2), pp. 158–177.

Welling, B. H. (2014) 'On the "inexplicable Magic of Cinema": Critical anthropomorphism, emotion, and the wildness of wildlife films' in von Mossner, A. W. (ed) *Moving Environments: Affect, Emotion, Ecology and Film*, Wilfred Laurier University Press, Ontario, pp. 81–102.

Wells, P. (2009) *The Animated Bestiary: Animals, Cartoons, and Culture*, Rutgers University Press, New Brunswick, NJ and New York.

Whitley, D. (2014) 'Animation, realism and the genre of nature' in von Mossner, A. W. (ed) *Moving Environments: Affect, Emotion, Ecology and Film*, Wilfred Laurier University Press, Ontario.

Wittgenstein, L. [1921] (2001) *Tractatus logico-philosophicus*, Routledge, London.

Wynne, C.D.L. (2007) 'What are animals? Why anthropomorphism is still not a scientific approach to behavior' in *Comparative Cognition and Behavior Review*, Vol. 2, pp. 125–135.

3   When animals look

Introduction

If anthropomorphism was expelled from the domain of 'serious science' in the early twentieth century, the same cannot be said for the popularisation of science where commercial imperatives, narrative appeals to audiences and the influence of Darwin's anecdotal writing style continued to shape naturalists' accounts of nonhuman animal lives. In periodicals and popular books, observations of ants were thought to confirm that highly stratified societies were natural, the gendered behaviours of insects were discussed in relation to human frameworks of social propriety, and much of the arthropod world could be recruited to provide moral exemplars for human conduct. In the twentieth century, film and later television inherited the legacies of earlier popularised observations of the 'natural world'. Since the 1980s, trends in US and UK natural history and wildlife television programming have accommodated certain commercial pressures by pursuing forms of genre hybridity (for instance drawing on soap opera and melodrama conventions) and spectacle (IMAX, 3D, and blue-chip forms) to attract audience numbers (Chris, 2006; Molloy, 2011). In these popular natural history and wildlife programmes, human social structures and the application of moral frameworks to nonhuman animal narratives have remained apparent. This chapter considers the mediation of arthropod experience by one such programme, the deployment of similitude in the form of gendered and moral norms, and whether or not a praying mantis can intervene in our thinking about ethical boundaries. Here I am interested in how questions of animal agency arise through alternative narratives of human failure to 'control' other animals during a production, and how this might open the possibility of empathetic imagining at a site of affective animal labour and commodified anthropomorphism.

Who is looking?

In *The Making of Micro Monsters*, a behind-the-scenes documentary that accompanied the six-part David Attenborough *Micro Monsters 3D* series, there is a moment of jubilation amongst the crew when a praying mantis turns her head abruptly to look straight into the camera with, what viewers are encouraged to interpret as, a menacing stare. This piece of footage is reused throughout the series. Her turn to camera is the final image in the trailer for the series and the iconic moment at the end of the title sequence for each episode, after which the words 'Micro Monsters with David Attenborough' appear. The dramatic percussive musical accompaniment which would be well-suited to a horror or science fiction movie, the proximity of her threatening turn towards the camera, her 'unblinking stare', the hostile position of her spiked raptorial forelegs, all of these visual and auditory cues are mobilised to support the overarching premise of the series: to use 'the latest 3-D technology to bring to life the extreme and deadly unseen world of bugs' (Atlantic Productions, n.d.). In *Micro Monsters*, a natural history programme, the praying mantis is made deadly through an accumulated history which precedes her and which the programme makers draw on to emphasise her iconic female monstrousness. What is centralised in this short mediated encounter is the look of a mantis; in other words, the returned gaze.

After John Berger (1980) and Jacques Derrida (2002), the look at and between humans and other animals has become significant as a means by which human/ nonhuman animal power relations are critiqued. Berger's classic essay 'Why Look at Animals?' centralised the role of 'the look' between humans and other animals and the importance of their visual representations. Anthropomorphism, Berger claimed, until the nineteenth century, expressed human proximity to nonhuman animals, but as those 'real' animals disappeared from our immediate experience anthropomorphism was subsumed into the visual encounter with representations of other species. As the distance between human and nonhuman animals grew, so did our unease with anthropomorphism until all that we were left with were inauthentic anthropomorphic relationships with animals – as pets and as visual spectacle. 'Animals are always the observed', Berger wrote (1980: 16). Thanks to capitalism, our relationships with real nonhuman animals were replaced with the inauthenticity of Disney, zoos, pets and toys, a situation that Berger judged to be 'irredeemable' (Berger, 1980: 28). Despite his verdict being pessimistic and full of gloom, in the late twentieth century, Berger's essay was important in that it opened up a debate about animals in the visual arts and culture.

Derrida considers the exchange of looks between human and nonhuman animal and specifically between a cat and himself. Instead of the nonhuman animal always being the observed, Derrida writes about an encounter in which a cat looks at him and he must respond. Crucially, the cat is not, Derrida reminds us, a fictional cat, 'the figure of a cat', but a real cat. What he describes is a material encounter, one which is not mediated by a camera or pen, one that reminds us that the mediated and unmediated look are discrete modalities of seeing. In the mediated encounter we are presented with the illusion of a returned gaze which is shaped by a confluence of discourses. How then are the look, the looking and the gaze intrinsically bound up with our understanding of nonhuman animals as objects or subjects, their similitude and difference to us, and with the attribution of subjectivity and the charge of anthropomorphism? It is arguably easier to work through such questions when we are thinking about those species we consider companions: dogs, cats, horses. What happens when we instead think about the praying mantis, her look and how she is mediated? In this case, I refer to not just any praying mantis but the one who appeared in the televised moment described above. What happens when we reconnect the symbolic with the material life of a mantis?

Why look at faces?

There is a well-documented fascination with the praying mantis amongst artists and scientists – a result, argued William L. Pressly about the Surrealists, of her 'exceptionally anthropomorphic form' and 'extraordinary mating rituals' (Pressly, 1973: 600). Writing about the praying mantis over a series of articles in the 1930s, Roger Caillois claimed that 'mankind has been highly struck by this insect. No doubt this results from some obscure sense of identification, encouraged by the mantis's remarkably anthropomorphic appearance' which, he claimed, 'seems to me an infallible source of its hold on human emotions' (Caillois, 2003: 73). In 1912, playwright and poet Maurice Maeterlinck described

> the ecstatic insect with the arms always raised in an attitude of supreme invo-cation, the horrible *Mantis religiosa* or Praying Mantis [. . .] she eats her husbands (for the insatiable creature sometimes consumes seven or eight in succession), while they strain her passionately to their heart.
>
> (Maeterlinck, 1912: xxvii)

Maeterlinck, like Caillois, was a keen admirer of entomologist Jean Henri Fabre who, in his discussion of the mantis, wrote primarily about the female insect and emphasised the discontinuity between her feminine appearance and behaviours. According to *Fabre's Book of Insects*, an early twentieth-century popularised translation of Fabre's *Souvenirs Entomologiques* (1879), the mantis was 'fierce as a tigress, cruel as an ogress' while possessing 'a certain beauty, with her slender, graceful figure, her pale-green colouring, and her long gauzy wings' (Fabre trans. Mattos, [1921] 2013: 24). Indeed, the *Micro Monsters* turn-to-camera shot could have been designed to illustrate Fabre's famous observation of the praying mantis: 'Having a flexible neck, she can move her head freely in all directions. She is the only insect that can direct her gaze wherever she will. She almost has a face' (ibid.). The affective power of her look was, according to Fabre, calculated to ter-rorise her victim: 'the Mantis stands motionless, with eyes fixed on her prey. If the Locust moves, the Mantis turns her head. The object of this performance is plain', contended Fabre. 'It is intended to strike terror into the heart of the victim, to paralyse it with fright before attacking it. The Mantis is pretending to be a ghost!' (Fabre, 2013, 26). In the programme's striking shot of a praying mantis, the inten-tion of *Micro Monsters* is unambiguous; the viewer is the locust and the construc-tion of the mantis' gaze is intended to kindle our deep-seated apprehensions about her. What is perhaps most troubling is that she does not turn away from our close observation. Instead she seems to look straight back at us, the pseudopupils of her large compound eyes a particular feature of her gaze. As Giovanni Aloi points out,

'Among insects, the praying mantis is able to return the gaze; and what a brave confrontation it is' (Aloi, 2011: 97).

Even though a praying mantis is able to return the gaze, Aloi suggests, we struggle to find the same kinds of response as we might have if the encounter were with another species, particularly a mammal (Aloi, 2011: 97). To what extent is that general discomfort with her gaze the result of, what Fabre referred to as, the mantis 'almost' having a face? In granting her agency and control over where and how she looks, Fabre's praying mantis is, he tells us, unique in the insect world, and in that regard more human-like; then he imaginatively transplants her gaze into the context of an incomplete or partial face. This articulation of the duality of similitude and difference is striking; here the subjective agency of the returned gaze is paired with the unsettling envisioning of an incomplete physiology. In Fabre's description, it is her gaze specifically that suggests the presence of a face, almost. In *Micro Monsters*, the framing of the praying mantis places emphasis on her two large compound eyes. The packaging of the *Micro Monsters* DVD collection continues the theme and plays further upon our desire for the returned gaze, with each of the four DVD covers sporting a close-up head shot of an arthropod who seems to look straight at us. In each case a pseudopupil and highlight are visible in the eye(s), the two forelimbs are in focus while the remainder of the body is edited, out of focus or strategically cropped to ensure that the humanlike visual cues outweigh those of the arthropod's alien-ness. As Attenborough's narration tells us at the beginning of each episode, these are 'creatures utterly unlike us; almost alien', and yet the series' narration and all of its marketing go to great lengths to engage the viewer in a world that is recognisable, known and understandable in western human terms – socially, morally and culturally. To these ends, the highlight in the eye of each arthropod is important, something Cheryce Kramer refers to as a sensory cue and a means by which the viewer is encouraged to identify 'a hint of consciousness', without which 'the eyes of most animals in wildlife photography look flat, dumb, or muted' (Kramer, 2005: 145). The pseudopupil, however, is illusory, an optical phenomenon that appears to the human spectator as a dark spot that seems to move across the insect eye as the observer moves. This gives the illusion of being watched intently by an arthropod. It is perhaps a narcissistic conceit of our own biology that leads us to think that we humans are the centre of another being's rapt attention. 'We speak of all manner of living things as having faces', Daniel Black writes, 'and even the crudest representation of key facial attributes – most importantly eyes – are enough to trigger recognition as a face' (Black, 2011: 1). Noa Steimatsky observes that 'the face sustains the gaze; it compels our attention and animates our responsiveness, our recognition' (2017: 6). If the return of the gaze and the face are coextensive, what does it mean to *almost* have a face, and in what ways does this shape the dynamics of our encounters with other animals?

Black proposes that our ability to recognise faces from only the most basic of representations 'suggest[s] that the face exists more in the mind of the viewer than on the body of the viewed: it perhaps results', he writes, 'more from the attribution of a face than the simple presence of physical features' (Black, 2011:

2). Humans are predisposed to see faces 'whether they are there or not', argues Stewart Guthrie (Guthrie, 1993: 105), because 'our preoccupation with a human prototype guides perception in daily life' (1993: 91). According to Guthrie, anthropomorphism is an outcome of perception which leads us to 'see faces in mountains, clouds, and automobiles' (1993: 105). 'Whether acquired or innate' he proposes, 'our models of faces [. . .] are powerful for good reason: "no other object in the visual world is quite so important to us"' (1993: 105). To be sure, it is of significance that four of the five human senses are located on and around the face. Our communicative and sensory experiences of the world confer high value to the face, so much so, claims David Le Breton, that

> in our societies, the principle of identity rests essentially on the face; to divest oneself of it through a mask, a veil, or painting of the face, is an act of great import by which the individual, sometimes without even knowing it, crosses the threshold of a possible metamorphosis.
>
> (Le Breton, 2015: 6)

So central is the face to society and culture that histories of the visual arts can be told through the changing depictions of the facial image. Writing about its aesthetic significance in the fine arts, Georg Simmel claimed 'in the features of the face the soul finds its clearest expression. [. . .] Aesthetically, there is no other part of the body whose wholeness can as easily be destroyed by the disfigurement of only one of its elements' (Simmel, 2004: 5). And Roland Barthes marks out an historical transition in cinema through the faces of Greta Garbo and Audrey Hepburn (Barthes, 1972: 55–56). Its ubiquity in contemporary media, 'in news broadcasts, presidential debates, interrogations and interviews, talk show confessions, Facebook, Facetime, Instagram' and cinema, however, leads Steimatsky to ask if the face has succumbed to 'promiscuous circulation to the point of exhaustion' (Steimatsky, 2017: 6). Certainly, the circulation of images of faces has never been so great, and it is this concentration of emphasis in popular culture that reflects the extent to which the face remains central to notions of self and identity. The face individualises and identifies us; portraits across various traditions of representation depict the face; photographic head shots are used for identification cards, passports and licenses. It also acts as more than just a representation of our outward appearance. In this regard, we continue to harbour expectations that the exterior image reveals the inner self. As Richard Twine argues, we 'retain the belief in a static correspondence between external image and morality and character' (Twine, 2002: 72).[1] As the language of cinema evolved in the twentieth century, film and particularly the close-up gave audiences a new means by which the face could be scrutinised and the relationship between external appearance and character organised. Trends in bodily enhancement now present an imperative to the contemporary consumer to beautify the outer self to match some notion of inner beauty and moral goodness. Where the standards of facial beauty are set by celebrities and stars, cosmetic surgery offers 'ordinary people' the opportunity to 'have a face like that of their heroines' (Featherstone, 2010: 203). The human face is a locus of signification and selfhood but without a fixed universal model or ideal; the face is individual, malleable, mobile, expressive, affective and communicative.

Although the most common image in western art is that of the human face, the first faces drawn by early humans were those of animals. Until the end of the Paleolithic period, where human forms did appear they had animal faces, partial animal faces or were depicted as faceless (Brener, 2000: 8). The first symbols were animals, claims John Berger (1980: 8), and the faces of those nonhuman animals were more detailed, their anatomy more accurately described than that of humans (Brener, 2000). Thereocephaly – having the head of a nonhuman animal – retains its symbolic weight, argues Erica Fudge in her analysis of Milton's representation of animal headedness. Transformed, animal-headed humans are de-individualised. 'They are', she argues, 'with their animal heads, *faceless*, because in this discourse a face is where the rationality that lies within is projected out into the world' (Fudge, 2013: 180). 'Lack of reason', Fudge proposes, 'means lack of face, means lack of individuality, lack of home, which in turn means that these beings are outside of full ethical consideration' (ibid.). In Philip K. Dick's post-apocalyptic envisioning of human-nonhuman boundaries and identity, *Do Androids Dream of Electric Sheep*, the subhumans or 'specials' are referred to as 'chickenheads' or 'antheads', a metaphoric thereocephalous being whose social subjugation is predicated on their lack of reason and inability to achieve the minimum pass in a mental faculties test (Dick, 1993: 19).

In Dick's novel, nonhuman animals and empathetic responses to their suffering are the measure by which humanness is decided using a Voigt-Kampff test – a series of questions and scenarios designed to elicit empathy in human subjects. Towards the end of the novel an android named Pris cuts off a spider's legs in a horrific act of mutilation. In the scene, the spider's difference is centralised and the qualification for their otherness is simply that they have eight legs. Pris asks the chickenhead character, J.R., why a creature would need so many legs and is dissatisfied that a spider should be eight-legged. The android logic is to remove four of the spider's legs to see what happens. The scene makes clear that it is their inability to feel empathy that enables Pris to experiment on the spider. Yet the removal of four of the spider's legs might also prompt a question of whether they are making the spider more familiar, perhaps more mammalian. Indeed, there is an uncomfortable sense in which this scene tests our empathy for the spider whose difference from us has been made so logically apparent. Outside the narrative world of the novel, empathy for four legs is much easier than empathy for eight legs. Pris eventually snips and cuts away four of the spider's legs while the androids talk about empathy and the possibility that it is merely an illusion, a way of separating them from humans. The spider survives, manages to creep away when one of the androids holds a lighted match nearby, and the narrative prompts us to identify with Isodore and the 'chickenhead's' empathy for the mutilated arachnid. Even as an animalised type of humanness, Isodore empathises with the spider's predicament and the reader is aligned with his affective response. Indeed, the reader is placed in the position of taking a version of the Voigt-Kampff test, and this scene makes appeals to our moral concern for a spider – a 'faceless' arthropod.

Fudge reminds us that the face is an important concept in human-nonhuman animal relationships in that it remains a marker of ethical status. From her reading of relationships in the seventeenth century, Fudge contends that 'to possess a face does not simply mean that an animal becomes a member of the community, it allows for levels of complexity – symbolic, ethical, economic, religious' (Fudge, 2013: 190). It is not the case, she argues, that having a face and therefore individuality automatically grants ethical status (2013: 195). She does however point out, through her reading of Donna Haraway, that to have a face is to be recognised as a subject and that this is a two-way street. In discussion of a specific material encounter with another animal, looking from their point of view enabled recognition of a 'subject with whom they could communicate' (Fudge, 2013: 181). 'Animals as well as humans are possessed of what might be termed faced-ness', Fudge remarks (ibid.). Emmanuel Levinas also argues that the face gives rise to consideration of our ethical obligations. Despite his claim that 'one cannot entirely refuse the face of an animal. It is via the face that one understands, for example, a dog' (Levinas, 2004: 49), Levinas asserts that a human face has priority over the face of a nonhuman animal. For Levinas 'the face' is not as a matter of biology but a form of ethical address. When it comes to whether or not particular nonhuman animals have the power to address us ethically, Levinas claims, 'I don't know if a snake has a face. I can't answer that question' (ibid.). Such issues are given a somewhat more blunt treatment in popular culture where the declaration 'I won't/ don't eat anything with a face' is an explicit claim to a moral position.

In visual representations, Aloi proposes that when confronted with certain forms of similitude, for example, when the eyes of a human are superimposed on the face of a nonhuman animal, 'it makes a demand for a different relational mode through the allusion to an emotional world similar to ours' (Aloi, 2012: 53). But, he points out, there are limits to this, and Aloi refers to the manipulated image of a snake by Nicky Coutts, saying that the image is 'more challenging' than those of 'warm-blooded animals' and he suggests that the artist does not go so far as to manipulate images of insects 'because, much more than other animals, insects do not have such a thing as a face. They are all-eyes, all-mandibles and antennae, I do not know if an insect has a face' (ibid.). Arthropods test out the limits of our discomfort with other animals; they confront us with our speciesist prejudices and ask us where we choose to draw our boundaries of difference. They are perhaps the easiest to ignore in abstract discussions of animality because, when it comes to nonhuman animals, our perception of degrees of similitude appears to relate to matters of ethical obligation. If, as Levinas and Aloi suggest, snakes test our limits, then arthropods are an even greater challenge for our moral sense.

Interested in what Merleau-Ponty has to contribute to a discussion of animal ontology, David Morris argues that faces are a particular feature of the animal world which 'makes explicit that seeing and being seen are not two separate things, but are part of one circuit of being' (Morris, 2007: 130). Morris describes the importance of the face to forms of reciprocity between humans and other animals, when he writes, 'An animal face is the face of a body, and expresses the whole of that body. This is vivid in the human case: *I face all of you in your face*' (2007: 132). For other animals as for humans, Morris contends,

> When I look in your face I don't just see you face, I see *you*, your feelings, your attention, a further whole of you, shining in your face. [. . .] I also see in dog, cow, deer, rabbit, horse, orangutan, parrotfish, octopus and stingray faces their seeing of me as something more than just surface.
>
> (2007: 133)

Encounters that involve seeing and being seen by other animals are then a mutual recognition of subjectivities. When he writes, 'I see this not just in the face in the usual sense, but in the animal body as a whole – the whole body serves as what I call a "greater face"', Morris offers a way to move beyond the humanistic view of the face by suggesting that it is embodied differently across species.

But in the mediated encounter we are not seen by other animals. The mediated animal experience is not a two-way ocular gesture. Our seeing is one-way and the returned gaze is illusory. There is only the imagined moment of cross-species recognition in the case of the mediated insect. When they can return the gaze and there is the potential for recognition of subjectivity, spiders, insects and other arthropods seldom trouble the boundaries that guard what are arbitrarily considered as the ethically worthy species. Mantises have five eyes, two of which are the large compound eyes that seem to return the gaze; the other three are simple eyes located in the middle of the head and below the antennae, which are used for smell. A mantis has one ear located in their ventral midline between the metathoracic legs, the purpose of which is to hear bat echolocation. The mantis' 'face' is located across her body and we cannot, as Thomas Nagel (1974) suggests, begin to understand how she experiences the world. Is it for this reason that in the ethical sense she is faceless? Could arthropods to be brought into the elite circle of ethical consideration that humans extend to some nonhuman others? Dick's novel imagines how an arthropod can, under certain conditions, appeal to our moral sense. We can be attentive to their situation even without ascribing them any degree of human similitude, instead emphasising aspects of their spidery difference. But what *Electric Sheep* also makes clear is that the context for this is situational, and relational in the sense that it depends on the lens through which our humanness is refracted back to us.

Why look at bodies?

Caillois, following Fabre, wrote about the various superstitions and beliefs that accumulated around the praying mantis. They were, Caillois speculated, the first insects on earth, and he repeated Fabre's claims that they could point the way home to lost children and that their nests were a cure for chilblains and toothache. Worshipped as a 'beneficent deity' and 'creator of the world' by the Khoikhoi and San peoples, Caillois wrote, within 'noncivilized' belief systems, 'what clearly seems to be most emphasized is the digestive dimension' (Caillois, 2003: 72–73). In this regard, Caillois argued that there was some 'coherence' to the collective human enthralment with praying mantises. Using examples of his immediate circle of peers and friends, Caillois stated, 'people even today are unambiguously drawn to the praying mantis' (Caillois, 2003: 76). He attributed the specific fascination with the mantis to two interrelated aspects: a reaction to their human-like form and a psychological response to the female's penchant for sexual cannibalism. With regard to the latter, Fabre wrote in *Social Life in the Insect World* (1912) a lengthy description of the courtship, with the male praying mantis as 'a slender and elegant lover' who 'makes eyes at his powerful companion; he turns his head towards her; he bows his neck and raises his thorax. His little pointed face almost seems to wear an expression' (Fabre, 1912: 82). And while he notes that 'insects can hardly be accused of sentimentality', Fabre's sympathies are clearly with the male when he writes, 'but to devour him during the act surpasses anything the most morbid mind could imagine' (Fabre, 1912: 84). Fabre's account of 'the act' is detailed: 'he is seized by his companion, who first gnaws through the back of his neck, according to use and wont, and then methodically devours him, mouthful by mouthful, leaving only the wings' (1912: 82). To his shock, Fabre one day finds a male

> in the performance of his vital functions, holding the female tightly embraced – but he had no head, no neck, scarcely a thorax! The female, her head turned over her shoulder was peacefully browsing on the remains of her lover! And the masculine remnant, firmly anchored, continued his duty!
>
> (Fabre, 1912: 84)

Similarly, absorbed by this particular aspect of mantis behaviour, Maeterlinck described the horror of the female's sexual proclivity: 'Her inconceivable kisses devour, not metaphorically, but in an appallingly real fashion, the ill-fated choice of her soul or her stomach' (Maeterlinck, 1912: xxvii). Caillois speculated that 'beheading the male before mating' might 'obtain a better and longer performance of the spasmodic coital movements, through the removal of the brain's inhibitory centres'. He hypothesised, 'it would hence be the pleasure principle that compels the female insect to murder her love' (Caillois, 2003: 78).

The conflation of sexual and nutritional appetites occupied many of the Surrealists. These were in Caillois' terms 'two savage desires' (Caillois, 2003: 77). In the first half of the twentieth century, Surrealist works contributed much to the mantis mythology. Many involved in the movement, including Caillois, Salvador Dali, André Masson, Alberto Giacometti and others, became fascinated with the cannibalistic desires that drove a female praying mantis to mate with an unsuspecting male then decapitate him and devour his body. Dali's *Cannibalism of the Praying Mantis of Lautreamont* (1934), Masson's *Landscape with Praying Mantis* (1939) and Giacometti's *Woman with Her Throat Cut* (1932), for instance, demonstrated the Surrealist fascination with this relationship between desire and death.[2] 'To them', writes Ruth Markus, 'she [the praying mantis] embodied the most negative female archetype, the "castrating woman" who represents cannibalism and death' (Markus, 2000: 33). In his account of the praying mantis' courtship, Fabre offered a tentative caveat to his observations, pointing out that they all took place under laboratory conditions. 'I might urge', he wrote 'in mitigation of these conjugal atrocities that the Mantis does not commit them when at liberty' (Fabre, 1912: 83). He later revises his thoughts after further observations to claim that the appetites of the female are a residue of their ancient legacy from the 'carboniferous period' before insect manners were gentle and when 'the lust to destroy' was the most beneficial way 'to produce'. 'I do not deny', wrote Fabre, 'that the limited area of the cage may favour the massacre of the males; but the cause of such butchering must be sought elsewhere' (Fabre, 1912: 85). Fabre's speculative pondering about the influence of the surroundings on the behaviour of the mantises was however later borne out by two studies which claimed that sexual cannibalism did not occur at every mating and the number of occasions were cited in 17 and 31 per cent of observations, respectively (Hurd et al., 1994; Lawrence, 1992). Despite such evidence, and although Attenborough, in the 'Making of' documentary briefly points out that the frequency of sexual cannibalism in the wild remains unknown, the substantive stories in the *Micro Monsters* episode continue in the tradition of the voracious female cannibal discourse. Indeed, Attenborough's narration of the praying mantis courtship would not be out of place in the pages of Fabre's account of the same. It is thus to the intersecting patriarchal and scientific discourses that construct the context of a site of commodified anthropomorphism and affective animal labour that this chapter now turns.

In *Micro Monsters*, David Attenborough's presence both on screen and as the voice-over narration functions to assure viewers of the scientific veracity of the programme. In this case Attenborough is also credited as writer and presenter. The series follows a similar structure to other programmes made in Attenborough's later years. Although acting as narrator throughout, Attenborough himself appears only briefly onscreen at a filming location at the beginning, midway and at the end of each episode. The opening narration with a short piece to camera establishes the aims of the series and its unique take on the subject matter, in this case enabled by macro and 3D technology, followed by an introduction to the topic of the particular episode. Midway through the episode and designed to appear after an advertising break and the recap narration, another short piece to camera places Attenborough in close proximity to one of the arthropod species who will feature in the second half of the programme. An appearance from Attenborough at the end of each episode brings together the final recap followed by an 'on the next' teaser that gives a glimpse of the dramatic action viewers can expect from the next episode. The arc of the series uses a traditional life cycle paradigm to frame the episodes, each of which includes seven or eight sequences that focus on one species per sequence. Each of the six programmes is therefore episodic in structure, a series of short stories united by an overarching theme of 'Conflict', 'Predator', 'Courtship', 'Reproduction', 'Family' and 'Colony'.

The praying mantis sequence is reserved for the end of the third programme – the most dramatic of the rituals featured in the 'Courtship' episode. *Micro Monsters* is in the expository documentary mode; it uses a rhetorical frame and addresses the viewer directly with a 'voice-of-authority commentary', where the speaker is both heard and seen (Nichols, 2010: 167).

Attenborough's presence as voice-over and on screen are vital aspects of the programme, providing the authoritative male voice that is traditionally associated with expository documentaries coupled with the (supposedly) inarguable expert professionalism of the celebrity naturalist. The narration for the praying mantis courtship sequence is heterodiegetic, a convention that allows the omniscient narrator to comment on what is directly observable from a particular ideological stance. In the case of the courtship episode and the praying mantis sequence, that positioning relies on a gendered discourse that assumes an overtly masculine viewpoint. This is naturalised through the narration of the episode. Each sequence that precedes that of the praying mantis focuses on a male arthropod with the majority of the factual information concentrated on *his* physiology, characteristics and behaviours. In this sense, the *Micro Monsters* narration assumes a masculinised discourse on animal courtship which has been produced and perpetuated in male-centred theories of animal behaviour. In doing so it normalises a human and patriarchal ordering of the realities of nonhuman animal experience and validates them as natural gendered behaviours.

Each sequence introduces us to the male arthropod first, a narrative strategy that encourages identification with his character in each 'story of courtship'. Moreover, the rituals described are ones in which the male is constructed as having the active role. One Goliath beetle fights with another male for the right to mate. We 'know he is a male', Attenborough states in his piece to camera, 'because he's got these two horns on his head with which he battles for females'. We are told about his strength; he can lift 'eight hundred and fifty times his own bodyweight'. The female doesn't need horns, the voice-over informs us; this fact accompanies the shot of a static female Goliath beetle who is moving only her mouthparts. Shots of stationary females and active, energetic males continue to normalise the gendered discourse of courtship. A male Chilean rose tarantula seeks out a female. He is shown moving through dense undergrowth and over tree branches. A female tarantula emerges from a small hollow and is, from then on, discursively passive as the narration informs us that 'he pushes himself beneath her', he uses 'special horns on his legs to prevent her biting him', his movements 'stimulate her' and 'with this she yields'. She is reduced to little more than a receptacle for 'offspring' which, we are informed, 'will all be his'. In the next sequence, the voice-over tells the viewer that the Nasonia male jewel wasp only lives two days so must busy himself with finding as many females as possible; the shots depict active males moving across the screen and inactive females waiting to be stimulated. 'He has no time to waste' on a female who is not receptive. A male Tanzanian red claw scorpion 'dances' with a female, he arouses her and although 'she is testing his strength', she eventually 'yields', at which he point he leads her over his sperm deposit. 'He's proved his strength and agility', Attenborough's voice-over claims, reinforcing what is important in a value-laden discourse of Darwinist competition. A male tramp ant must rid the nest of all other males to secure his right to mate, and other than one shot of the female ants to illustrate how to tell them apart – the sexually ready female ants have wings – the rest of the sequence focuses on the battle between males and the strategies used by a dominant male to secure 'rights' to mate. In a sequence about tropical house crickets, the male crickets, although not moving around, are seen to be doing all work of court-ship, rubbing their legs together to attract a choosy female. Central to this story is the male's strategy to keep a female cricket occupied. He produces a parcel of sperm and syrup described as a 'present'; she eats the syrup which 'distracts her while fertilization takes place'. The notion of the male securing his 'line' is again reinforced; he 'fathers her offspring'. Throughout these sequences, the narration constructs females as inactive, passive (and, if not quite passive then dominated until they 'yield' to the male), choosy (wasteful of the male's time), reduced to carriers of the male's offspring and distracted by nutritional desires. The males, always introduced first in the sequences and given substantially more screen time than the female arthropods, are by contrast active, combative, dominant, strong, competitive, focused and capable of creating and deploying strategies that enable them to reproduce.

The final arthropod sequence moves on to the praying mantis' mythologised aberrant sexual behaviour which the narration sets up for the viewer in a radically different way to the previous sequences. The voice-over informs us that the next insect 'symbolises the extremes that some animals will go to in order to mate'. What follows is a series of clips of praying mantises catching other insects and consuming them, with the gory bone-crushing sound effects and urgent music required to drive home the message that these are 'highly skilled predators', 'vora-cious', able to catch and consume other creatures 'larger than themselves'. With the rapacious appetite of the praying mantis established, the narration introduces the male mantis as he climbs up a stick towards a 'potential mate'. The sequence cuts to a shot of the female mantis at the top of the stick swaying from side to side; the voice-over gives us no clue as to why she might be moving in this way. In the next shot the female reaches down and catches the male; Attenborough informs us that she may have been 'put off by unwelcome sexual advances or driven by hunger'; the result however is that she 'begins to eat her suitor'. The narration is purposefully graphic about the stages of her consumption of the male's body. She removes 'his head and the brain cells that control his inhibitions'. The sequence echoes Fabre's horrified observation of the decapitated male continuing to mate with the female mantis. After the male mantis' head falls to the ground, Attenbor-ough confirms that 'he's begun to impregnate her', and at the end of the sequence he assures the viewer that 'in reproductive terms this male has succeeded'. 'But', he cautions, 'his death is a symbol of how strangely unfeeling the arthropod world can be'. Although written more than a hundred years apart, Attenborough's char-acterisation of the event as 'strangely unfeeling' shares discursive resonances with Fabre's pity for the male mantis and the lack of insect sentimentality. In *Micro Monsters*, female arthropods have already been established as distracted by their drive to eat, the framing of the mantis emphasises how the consuming female body is cruelly indifferent and voracious with interwoven sexual and gluttonous desires. From the patriarchal viewpoint that the narration assumes, she is an unfeeling body without the ability to regulate her desires while he, on the other hand, is a suitor with a brain capable of controlling inhibitions who, even in death, is successful. In the insect world of *Micro Monsters*, the anthropomorphic narration decontextualises nonhuman animal difference and offers it back to audiences as an overplayed discourse of gender difference. The audience is encouraged to understand these nonhuman animals through an abstraction of human social relations.

While it is inevitable that a programme such as this will adopt a popular discourse to ensure that the subject matter is appealing to the audience, this patriarchal anthropomorphism plays on gendered assumptions that are validated as 'natural' by the framing of nonhuman animal behaviours. It might be the case that the (male) viewpoint makes affective appeals to the viewer but these are based on the anthropocentric moral framing of the encounter as 'unfeeling'. In a previous sequence where the dominant male tramp ant has successfully engineered the dismemberment of another male ant, we are told that in this situation, 'violence pays'. In this way, male-to-male 'violence' is naturalised through the gendered Darwinian contest as necessary and even noble, while what is constructed as 'female to male aggression' is conceived of as lacking in emotion, strange and emblematic of extremes of behaviour. Throughout the 'Courtship' episode, the female body is constructed as a primarily static object to be impregnated or mated. When she is anything more than this, her 'feminine' character is brought into question. For most of the episode, female arthropods are denied agency while the males are granted a fuller and more complex characterisation, albeit based on the tropes of a traditional form of hegemonic masculinity and a 'survival of the fittest' discourse. In this context, similitude and difference is articulated in service to anthropocentric and patriarchal notions about male and female bodies and any affective appeals rely on reductionist gender norms from behind the thin veil of scientific respectability that natural history programmes manage to maintain.

Caillois argued in his essays that the praying mantis had 'objective lyrical value' and even without knowledge of the scientific studies and myths that had accumulated around her, the mantis would 'disturb the affectivity of so many different types of people' (Caillois, 2003: 80). The mantis was, he proposed, an allegorical object that acted directly on the unconscious in such a way as to inspire fear and attraction – to such a degree, indeed, that scientists set aside their 'scientific detachment' to regard the mantis as a 'murderous mistress' or female android (Caillois, 2003: 78). Her lyrical value was heightened, Caillois suggested, by her anthropomorphic likeness to a female android – 'artificial, mechanical, inanimate and unconscious machine-woman' – that evoked a premonition of the relationship between love and death (ibid.). She holds her front legs up and together as if 'in prayer', the rationale behind her naming we are told across various accounts, but this serves to verify her deceptiveness. Her posture, which suggests an illusory bipedalism, may imply devout piety but is instead the stance of a predatory monster. 'Those pious airs', Fabre declared, 'are a fraud: those arms raised in prayer are really the most horrible weapons, which slay whatever passes within reach' (Fabre, 2013: 24). Fabre also wrote of the praying mantis in mechanistic terms, a description that lent itself well to the trope of the bad mother who Fabre noted had no interest or tender feelings for her family: 'The Mantis, I fear, has no heart. She eats her husband and deserts her children' (Fabre, 2013: 29). Fabre's and Caillois' descriptions anticipated the later cultural fascination with the mantis' various iterations as the femme fatale, the alien, and the monstrous female in film. All of this mythology is leveraged by the *Micro Monsters* programme makers, keen to give the audience a natural history documentary version of the monstrous horror that the series title promises. Technologically enabled to assume the physical viewpoint of an insect, the *Micro Monster* viewer is invited to look at a praying mantis who is visually amplified to fill the screen and rendered in 3D for the large screen television experience. In doing so, the narration constructs an anthropomorphic gaze under which the praying mantis is a hostile alien body – a deceptive, morally compromised castrating female. This mediated encounter is constructed in such a way as to amalgamate multiple types of visual encounter into the assumed authoritative objective look of the popularised science television camera. In this case, the observation of the professional naturalist directs the audience's 3D-enhanced experience of the 'viewpoint' of an arthropod.

From body to faced-ness

For many of us, the majority of our encounters with other species are mediated. We experience a screen-based menagerie as a confluence of the social, cultural, institutional, political, technological and embodied. In the case of the female praying mantis in *Micro Monsters*, the mediation of her embodied lived experience is governed, amongst other things, by the norms of human-insect relations; the accumulated histories of praying mantises (derived through knowledge forms that include the scientific and popular); interrelationships of discourses of gender, age, and other social divisions; televisual codes and conventions; institutional economics and practices; and the dynamics of production, affect and consumption. In other words, when we 'look' at nonhuman animals we engage with much more than 'the seen' and – if we acknowledge the importance of sound in audio visual media – 'the heard'. In *Micro Monsters*, the complexity of the praying mantis' being is mediated and viewed as a human embodied experience. In the moment when we look at her and she returns the look, her embodied experience is diminished by that mediation, and even with a nod to Nagel's (1974) argument that we do not have the perceptual capacity to ever understand her experience of the world, the construction of that moment is designed to make appeals to *our* feelings of discomfort. There is no sense in which this representation makes any empathetic appeals to the viewer outside of the gendered discourse that constructs the male as the sacrificial although still successful mate. Instead, the praying mantis is trapped in an institutional anthropomorphic gaze that has little concern with the realities of her life but is instead motivated by the need to provoke in the audience the affective response promised by the series title. She is a micro monster and must be imagined as monstrous with all the attendant registers of discomfort that such appellations should satisfy. Hiding behind the discourse of science and the combined authorial and institutional authorities of David Attenborough (the narrator), Sky (the broadcaster) and natural history programming (the genre), this particular envisioning of a praying mantis tells us something about her, and plenty about us. She is purposefully cast in her mythical role as the femme fatale while the conceit of the programme – that macroscopic technologies allow humans to enter the previously unseen world of insect-monsters – enlarges her gruesome presence on screen. Similitude and difference are called upon to compound her gendered monstrousness. She is made to be strangely familiar: she is like the duplicitous immoral human femme fatale of popular culture and embodies both human postural similarity and alien difference.

In addition to the discursive-affective configurations of mediation, our encounters with nonhuman animals in film and television and other screen-based media are also caught up in dynamic relations with paratexts, those ancillary materials – official and unofficial – that can frame the viewing experience. 'Making of' programmes, trailers, reviews, internet discussion and other textual proliferation create an expanded presence which might include narrative extension, amplification of particular aspects of a text or a change to the nature of its address (Gray, 2010: 1–30). Across this textual proliferation, the process of mediation is governed primarily by human concerns in the sense that popular media content is created by and for humans and utilises human frameworks; ideological, discursive, economic, social, cultural and so forth. Popular media is materially and discursively human-centred and narratives, whether fiction or non-fiction, ascribe some degree of intentionality and agency to human and nonhuman animal characters. If the charge of anthropomorphism rests on the denial of intentionality and subjectivity to nonhuman animals then popular media, which utilises 'everyday' rather than distancing mechanomorphic language (Crist, 1999), cannot avoid the anthropomorphic alignment of humans with other animals. Moreover, the peculiar character of our mass communication is species-specific such that it imposes, through the processes of mediation, a uniquely humanised configuration of all that falls within its purview.

Jonathan Gray points out that in the case of television creators, 'the paratexts of interviews, podcasts, DVD bonus materials, and making-of specials are their preferred means of speaking [. . .] as they will use paratexts to assert authority and to maintain the role of author' (Gray, 2010: 110). However, he contends, the proliferation of paratexts, including audience generated paratexts, can compete with this authorial authority to the extent that there is no 'easy predictability of outcome' (Gray, 2010: 111). This then offers critical animal studies a way to read texts against the grain – a resistant reading with the intention of reconnecting the symbolic to the material. In other words, paratexts can enable a resistant reading that acknowledges that the nonhuman animal who appears on screen is or was a living being with agency. This strategy recognises the material existence of a non-human animal and does not consign them to being only of value in a metaphoric or symbolic sense. By thinking of animals only in terms of their representations, we run the risk of erasing their individuality and seeing their agency as tied only to the affective resonances that they have through the text. This elevates the text and forces us to focus on the narration, the representation and so forth at the expense of the real nonhuman animal. When we read only the representation of a nonhuman animal without acknowledging that individual's material existence, we diminish, even erase, their being in the world and ignore the industrial and commercial realities that govern how they come to appear onscreen. Resistant readings across paratexts can suggest an altogether different narrative, one in which we might find unexpected empathetic connections while shifting the focalisation of a narration to look *with* rather than *at* other animals.[3]

In the case of the female and male praying mantises who appeared on screen in the *Micro Monsters* 'Courtship' episode, to read their story against the intended meaning is reasonably straightforward. One reason for this is that unlike much film, television and other media which rely on real nonhuman animals and the apparatus used to control them staying invisible to the audience, in the case of *Micro Monsters*, the difficulties of filming arthropods are a key aspect of the contextualising narrative that the programme favoured. This emphasises how arthropods have little moral value, especially when weighed against mammals, and for this reason there is little risk to corporate reputation in showing how the conflicts and 'predator/prey' interactions were staged and managed for the purposes of the production. Indeed, it was vital to the discourse of macro 3D technological innovation that the paratextual narrative showed how such stage management was necessary to achieve the spectacular experience of the arthropod viewpoint.

The *Making of Micro Monsters* documentary and press articles in advance of the broadcast produced a wraparound or framing narrative which (typical for natural history programmes) emphasised the technological advances needed to produce the series, the unique 'never before seen' footage and the difficulties involved in obtaining that footage. The series and its paratexts were quite open about the extent to which the setting and action were managed for the camera. One article dramatised the 'challenge' of

> persuading an emerald wasp to sting an American cockroach in a six-inch square area, a process which sees the cockroach become a 'zombified' host to a parasitic wasp which then feasts in its victim before emerging as a young adult.
>
> (Briggs, 2013)

Elsewhere it was suggested that arthropods were a perfect subject for a 3D television series as they are able to elicit an affective response, a mix of 'fascination' and 'a shudder' (Goss, 2013). Moreover, it was suggested that as the 3D experience involved getting closer to the subject; the article claimed, 'that's much easier when they're small and relatively oblivious to humans than it is with larger animals who might see the cameraman as a threat, or even dinner' (ibid.). However, the article noted that the studio in which the parasitic wasp and cockroach were filmed had 'hot lights, lots of noise and people' and that it took ten hours to finally get the shot they wanted (ibid.). It is this lengthy timescale, the inability of humans to 'get the shot' that reveals the agency of individual animals who do not comply with the wants and wishes of the programme makers or the narrative.

The reality of the experience with the praying mantises was similar to that of the wasp and cockroach – a lengthy period of filming under artificial conditions which viewers are told in the context of the difficulties of achieving the shots were inappropriate for the insects. For the series the male and female mantises were supplied by an 'insect wrangler' who also had responsibility to manage from the individual insects the behaviours that the narrative demanded. The mantises came from a private collection housed in a large garden shed in Norfolk in the UK. The female mantis was brought to the studio space in a plastic box and then placed at the top of a stick under the harsh lights needed for 3D filming. The insect wrangler placed the male mantis lower on the stick and warned the crew to stay very still. At one point in the 'making of' documentary, the audience is informed 'every time you touch it, you stress it' and when trying to create the 'notorious piece of insect behaviour', that 'one disturbance from the crew will turn the female into a killing machine'. The insect wrangler discussing the issues points out in the documentary that placing the mantises on a stick, having bright lights, cameras, movement and vibrations create inappropriate conditions for the insects who usually mate at dusk. It is clear from the documentary that it was always the intention to capture footage of the male being eaten by the female mantis. The encounter was stage-managed and intentional; the male and female mantis were put under artificial and extreme conditions to realise on camera the sexual cannibalism mythology. That the mantises did not immediately comply, that the shot took so long to achieve, that their management was teetering for so long at the brink of human failure – these are the resonances of the mantises' agency. The narrative of their performance and the capturing of that managed death was eventually celebrated as a 'success', a triumph of human knowledge and its application in the wrangling of the mantises and a feat of innovative technological prowess.

Conclusion

When we know about the realities of a nonhuman animal's experience, how then do we look at a narrative of their lives? What type of gaze do we bring to that viewing experience, and is it possible to usurp the discursive control of the authorised narrative with a revised narrative that places the other animal's experience, inasmuch as we can understand it, at the centre? In the case of a praying mantis on the set of *Micro Monsters*, it is possible to re-watch the 'Courtship' sequence and see the mantises from a different viewpoint, one that does not depend on a pseudo-Freudian gendered discourse of oral fixations but instead is contextualised by the lived reality of the mantises. When we know what happened to them, how the sequence was achieved, where their agency can be located and how their encounter and death was managed, does this make any difference to the way in which we 'see'? It is perhaps much easier to think about a similar situation that occurred before the release of *A Dog's Purpose* (2017), which was postponed after video footage from the set was made public. The footage was alleged to show a handler forcing a terrified dog into fast-moving water. The opening of the film was postponed because of the risk to box office profits and reputational credibility. Audiences would, it was anticipated, either boycott the screenings or would not be able to watch the scene as it was intended. Despite the emotional appeals of the film, the risk was that the scene would trigger the memory of the video footage and audiences would see a scared dog being mistreated and pushed rather than a brave dog jumping of his own volition in to the water. In this case, there is greater popular knowledge about dogs; we believe that we can read a dog and make sense of how she is feeling in that situation. Is the same possible in the case of an arthropod, or do insects need to be made into the cartoon bipedal, clothes-wearing representations of *A Bug's Life* or *Antz* before we can grant them some degree of consideration, moral or otherwise? Is there subjectivity in the look of the real praying mantis or is the gap too wide, forcing us to find it only in animated zoomorphic figures? This would necessitate a shift in focalisation from the intended 'insect viewpoint' to a viewpoint that is empathetic, in the sense that it moves between an imagined mantis perspective contextualised by the knowledge made available about the mantises on the set and mantises generally, and a third-person viewpoint which is constructed by the camera – in other words, by shifting the focalisation such that the anthropomorphic gaze (as if the viewer were the size of an insect) of the series is displaced as the primary viewpoint and takes on a different narrative role when contextualised by the imagined viewpoint that reading across the paratexts can give. The imagined viewpoint is not governed by 'irrational emotion' but by a different type of anthropomorphic envisioning of the insect's experience. In this way, it is possible to re-see the courtship sequence in another modality, contextualised by a different narration of the mantises' lives. In other words, this shifts the focalisation through an engagement with the lived reality of a nonhuman animal's life, connecting the material with the symbolic in such a way as to resist the gendered assumptions that are played out by the mediation of the mantises' bodies and instead acknowledging that the realities of cultural production involve the affective labour and performance of nonhuman animal others.

Notes

1 According to the Uncanny Valley hypothesis, the face also plays a critical role in our empathetic responses to others. While humans may, as Guthrie and Black suggest, have a particular propensity for seeking out human facial features, likeness and particularly a 'lifelike' face, might bring about feelings of discomfort to the point of revulsion if we take a slip into the uncanny.
2 Other artists whose works were inspired by the mantis include Pablo Picasso, M.C. Escher and Oscar Domínguez.
3 I have endeavoured to do this previously in *Popular Media and Animals*, where I give an account of the lives of Bonzo (a chimpanzee) and Fagan (a lion), who were constructed as Hollywood stars.

References

Aloi, G. (2011) *Art and Animals*, I B Tauris, London and New York.
Aloi, G. (2012) 'Beyond the pain principle' in Blake, C., Molloy, C., and Shakespeare, S. (eds) *Beyond Human: From Animality to Transhumanism*, Continuum, London and New York.

Atlantic Productions (n.d.) 'Micro monsters 3-D' online at www.atlanticproductions.tv/productions/micro-monsters/

Barthes, R. (1972) *Mythologies*, Paladin, Herts.

Berger, J. (1980) *About Looking*, Random House, New York.

Black, D. (2011) 'What is a face' in *Body & Society*, Vol. 17 (4), pp. 1–25.

Brener, M. E. (2000) *Faces: The Changing Look of Humankind*, University Press of America, New York and Oxford.

Briggs, S. (2013) 'Creepy crawly alert: Norfolk insect wrangler puts bug collection on show in new David Attenborough series' *Norwich Evening News* 16 June, online at www.eveningnews24.co.uk/views/creepy-crawly-alert-norfolk-insect-wrangler-puts-bug-collection-on-show-in-new-david-attenborough-series-1-2238278

Caillois, R. (2003) *The Edge of Surrealism: A Roger Caillois Reader*, Duke University Press, Durham and London.

Chris, C. (2006) *Watching Wildlife*, University of Minnesota Press, Minneapolis and London.

Crist, E. (1999) *Images of Animals: Anthropomorphism and Animal Mind*, Temple University Press, Philadelphia.

Derrida, J. (2002) 'The animal that therefore I am (More to follow) [trans. David Wills] *Critical Inquiry*, Vol. 28 (2), Winter, pp. 369–418.

Dick, P. K. (1993) *Do Androids Dream of Electric Sheep*, HarperCollins, London.

Fabre, J. H. (1912) *The Life of The Spider*, Hodder & Stoughton, London.

Fabre, J. H. (2013) *Fabre's Book of Insects*, Dover Publications, New York.

Featherstone, M. (2010) 'Body, image and affect in consumer culture' in *Body & Society*, Vol. 16 (1), pp. 193–221.

Fudge, E. (2013) 'The animal face of early modern England' in *Theory, Culture and Society*, Vol. 30 (7–8), pp. 177–198.

Goss, P. (2013) 'Are insects the unlikely saviours of 3D?' *Techradar*, 4 June, online at www.techradar.com/uk/news/television/can-insects-help-3d-crawl-its-way-back-into-our-hearts-1156136

Gray, J. (2010) *Show Sold Separately: Promos, Spoilers and other Media Paratexts*, New York University Press, New York.

Guthrie, S. (1993) *Faces in The Clouds: A New Theory of Religion*, Oxford University Press, Oxford.

Hurd, L. E., Eisenberg, R. M., Fagan, F. W., Tilmon, K. J., Snyder, W. E., Vandersall, K. S., Datz, S. G., and Welch, J. D. (1994) Cannibalism reverses male-biased ratio in adult mantids: Female strategy against food limitation? in *Oikos*, Vol. 69, pp. 193–198.

Kramer, C. (2005) 'Digital beasts as visual Esperanto: Getty images and the colonization of sight' in Daston, L. and Mitman, G. (eds) *Thinking with Animals: New Perspectives on Anthropomorphism*, Columbia University Press, New York.

Lawrence, S. E. (1992) 'Sexual cannibalism in the praying mantid, *mantis religiosa*: A field study' in *Animal Behaviour*, Vol. 43 (4), pp. 569–583.

Le Breton (2015) 'From disfigurement to facial transplant: Identity insights' in *Body & Society*, Vol. 21 (4), pp. 3–23.

Levinas, E. (2004) 'The name of a dog, or natural rights' in Atterton, P. and Calarco, M. (eds) *Animal Philosophy: Ethics and Identity*, Continuum, London and New York.

Maeterlinck, M. (1912) 'The insect's homer' in Fabre, J. H. (ed) *The Life of The Spider*, Hodder & Stoughton, London, pp. i–xxxiv.

Markus, R. (2000) 'Surrealism's praying mantis and castrating woman' in *Woman's Art Journal*, Vol. 21 (1), pp. 33–39.

Molloy, C. (2011) *Popular Media and Animals*, Palgrave Macmillan, Basingstoke.

Morris, D. (2007) 'Faces and invisible of the visible: Towards an animal ontology' in *PhaenEx*, Vol. 2 (2), pp. 124–169.

Nagel, T. (1974) 'What is it like to be a bat?' in *The Philosophical Review* LXXXIII, Vol. 4, October, pp. 435–450.

Nichols, B. (2010) *Introduction to Documentary* (Second edition), Indiana University Press, Bloomington.

Pressly, W. S. (1973) 'The praying mantis in surrealist art' in *The Art Bulletin*, Vol. 55 (4), pp. 600–615.

Simmel, G. (2004) 'The aesthetic significance of the face' in Blaikie, A. (ed) *The Body Volume IV: Living and Dying Bodies*, Routledge, London and New York, pp. 5–9.

Steimatsky, N. (2017) *The Face on Film*, Oxford University Press, Oxford.

Twine, R. (2002) 'Physiognomy, phrenology and the temporality of the body' in *Body & Society*, Vol. 8 (1), pp. 67–88.

4　When animals feel

Introduction

If the attribution of emotional states to animals other than humans was, for science, contentious during the twentieth century, popular culture naturalised the emotions of certain species to the extent that the emotional ranges of some animals and their signifiers have been established as common sense within the public consciousness. We can think for example of the wagging tail of a dog connoting happiness which, despite being a problematic over-simplification of this highly complex aspect of canine communication, has nonetheless become a widely circulated piece of popular knowledge that equates some outward behaviour with an internal emotional state for a species commonly regarded as a companion. In this chapter I explore how despite an abundance of popular representations of nonhuman animals expressing emotional states coextensive with those of humans, the negotiation and regulation of nonhuman animal emotion remains tied in many regards to the requirements of capitalism. Emotion is contextualised by welfare, scientific and other discourses that broadly situate companion animals as having emotional capacities that humans are responsible for maintaining, often through purchases of food, toys, treats and so forth. Farmed animals are depicted in marketing for animal products as having emotionally fulfilled lives within the context of a welfare discourse, while the realities of their lives and deaths within the animal agriculture industries are seldom accessible to the public. The emotional capacities of free-roaming (wild) animals are ambiguous, species-dependent and contextually driven by various forces including conservation, welfare and scientific discourse. Examining the politics of the mediation of animals' lives, this chapter explores the regulation of animal emotion at different sites of anthropomorphism. It looks at the management of anthropomorphism and how the authority of interpretation is constructed and the importance of gender and genre to its authorisation or disavowal.

Grieving kangaroos

In January 2016, images of a male kangaroo, a dying female kangaroo and a joey went viral on social media. The photographs were taken by amateur photographer

Evan Switzer on 12 January 2016 while he walked through a neighbour's property in Queensland, Australia. Switzer shared the images on social media, posting one photograph to the Facebook page of the *Fraser Coast Chronicle* with the comment:

> I noticed one was dead and as I kept walking I noticed the male kangaroo trying to pick up the dead female. I went back home and got the camera and sat and watched something truly amazing with the male and joey kangaroo mourning the loss of the female.
>
> (Switzer, 2016)

The *Fraser Coast Chronicle* (Formosa, 2016) featured the story on the front page of its print version on 13 January; *The Courier Mail* (Brisbane) followed suit on 14 January with the headline 'Tender-roo'. Throughout 13 January other media outlets in Australia, the UK and US picked up the story and reproduced the images in online articles with headlines such as 'I'll love roo till the end: Dying kangaroo mum reaches out to her joey in final moments' (Fruen, 2016); 'Photographer captures moment kangaroo cradles dying companion as joey looks on' (Khomami, 2016); 'Please don't die, mum: The heart-wrenching moment a mother kangaroo reaches for her joey one last time – before dying in the arms of her male companion' (Peters, 2016); and 'Heartbreaking pictures show mother kangaroo reaching for joey one last time before dying in male companion's arms' (Horton, 2016). Each article included a quote from the photographer describing the encounter: 'He would lift her up and she wouldn't stand she'd just fall to the ground, he'd nudge her, stand besides her . . . it was a pretty special thing, he was just mourning the loss of his mate' (Khomami, 2016).

Between 12 and 18 January Google searches from the UK and Australia for 'kangaroo', 'kangaroo dying', kangaroo mourning' and 'kangaroo photo' peaked in frequency on 14 January (Google Trends). The story received less media coverage in North America than in the UK and Australia. Nonetheless, Google searches for 'grieving kangaroo' and 'mourning kangaroo' showed peaks in frequency on 14 and 16 January respectively (Google Trends). Throughout 14 and 15 January, follow-up online media reports about the photographs included statements from a principal research scientist at the Australian Museum and a senior lecturer in veterinary pathology. These statements directly refuted Switzer's version of his encounter with the kangaroos. Media outlets reported one or both of the scientist's explanations:

> The male is clearly highly stressed and agitated, his forearms are very wet from him licking himself to cool down. He is also sexually aroused: the evidence is here sticking out from behind the scrotum (yes, in marsupials the penis is located behind the scrotum).
>
> (Eldridge, quoted in Gage, 2016)

'Interpreting the male's actions as being based on care for the welfare of the female or the joey is a gross misunderstanding, so much so that the male might have actually caused the death of the female' (Spielman, quoted in Hunt, 2016). *National Geographic* commented online about the story concluding, 'It appears this is a classic case of anthropomorphism. While it's in our nature to assign human feelings and behaviours to animals, the truth tends to be much more scientific – and much less cute' (National Geographic, 2016). Other media outlets which referred to the popular interpretation of the images as 'naïve anthropomorphism', ran with headlines that included 'Kangaroo in "grieving" photos may have been killed while trying to mate, scientist says' (Hunt, 2016), 'Humans apparently mistook kangaroo necrophilia for tenderness' (*CBS News*, 2016) and '"Grieving" kangaroo may have actually been a horny killer' (*New York Post*, 2016). One popular UK media outlet headlined the follow-up article: 'No, this kangaroo wasn't "grieving" – it was raping a dead female, experts say' (Readhead, 2016).

In press coverage of this story, anthropomorphism emerged most obviously as an object of discourse in the scientific criticism of the public interpretations of the photographic images. In this case, however, this is more than an account of how scientific discourse undermines a sentimentalised public understanding of kangaroos. Instead it exemplifies how human gendered norms are continually and routinely imposed onto other animals, how animal agency is discursively managed by those norms, how empathy is regulated in the public sphere and how that regulation is entangled with mediation of emotion and the act of bearing witness to animal death. Here the contest between popular and scientific knowledge was mediated by a press discourse that initially authorised the legitimacy of the photographer's encounter with the kangaroos. Switzer's description of the interactions between the kangaroos and public endorsement of his account, indexed by the viral status of the images, coalesced within the media reports to reinforce the veracity of the images' meanings, as evidence of human witness to kangaroo emotion and to the death of a female kangaroo. Only the first report in the *Fraser Coast Chronicle* offered another commentary on the images via a link to an article with the headline 'EXPERT: Kangaroos share bond with family'. In this piece, the curator of the Fraser Coast Wildlife Sanctuary was reported to have responded to the images of the kangaroos saying, 'it was a common sight to see animals in the wild displaying a nurturing attitude towards their family' and that he had witnessed similar bonding behaviours in wild birds and dingoes (*Fraser Coast Chronicle*, 2016). In reports from other media outlets, the photographer's commentary provided the only authoritative voice where he was positioned as witness to the (photographed) death of the female kangaroo.

The construction of the photographer as witness is important here in terms of thinking about the emotional appeals of the images and their meanings. Although the images were shared on social media prior to the press reports, these were not without context or reference to the photographer's comments. The images were then remediated as print and online news articles, the latter of which were widely shared via Facebook and Twitter. For example, the initial article in *Guardian Online* was shared over 92,000 times. By way of a comparison, the follow-up 'debunking' article had 10,700 shares. Across the first set of articles the photographer's status as amateur or professional was ambiguous. Described in the *Fraser*

Coast Chronicle first as a 'photographic enthusiast', articles in the *Guardian, New York Post* and *Brisbane Times* referred to him as 'a photographer' with, for instance, the *Guardian* mentioning only at the end of the article his attending a camera club that night as an inference about any amateur status. In this case, establishing the status of a photographer as having a claim to knowledge of the truth of events is intrinsically tied to the ways in which digital media practices intersect with contemporary understanding of what it means to be 'witness' to an event. Mette Mortensen argues that 'negotiations of the truth in connection with media representations of events in the material world often evolve around the figure of the eyewitness and the text of the testimony' (Mortensen, 2014: 12). Digital media enables eyewitnesses to create and distribute media content themselves such that they no longer simply 'make appearances in the media as sources of information'; they are integrated into the 'logics of the current media system' where the 'citizen photographer is often regarded as a successor to the traditional figure of the eyewitness' (Mortensen, 2014: 12).

The active sourcing of content from those outside the media industries has become a feature of contemporary news organisation practice where the rise of citizen journalism has emerged against a backdrop of budgetary cuts within the industry. Allan and Thorsen observe that while mainstream news media organisations 'decline to surrender their traditional editorial control, agenda-setting functions or gate-keeping authority', 'citizen journalism variously enters into and informs today's world news ecology with its overlapping formations and flows of news, mainstream and alternative news media, and new interactive technologies of news dissemination and user generated content' (Allan and Thorsen, 2009: xi). This has led to a blurring of the distinctions between amateur and professional which has divided academic and industry opinion on whether it represents a democratisation of journalism or a threat to journalistic standards. On social media, public responses to the photographs echoed these arguments with some users asking why reporters had declined to 'fact check' the initial interpretation of the images with an expert. Aside from the increasing currency afforded to the notion of a 'post-fact world' and the role played by news media in that world, a story about a dying kangaroo and animal emotion would be considered by mainstream media organisations as located at the far end of the 'soft' news spectrum and not requiring of the time and effort of journalists to verify the accuracy of the witness account or obtain expert opinion to qualify its legitimacy. While some public comments that followed the 'debunking' articles expressed dissatisfaction with what was perceived as a lack of journalistic rigour, the tone and focus of the press coverage reveals the all too familiar ways in which animals generally, and particularly when they are individuated, are regarded as trivial 'filler' stories. In an era of 24-hour news, filler stories – short articles of low or secondary importance – are used extensively to plug the 'news hole'. Due to their brevity, use as clickbait and as an entertaining digression from 'serious' news, animal filler stories often employ species stereotypes and perpetuate perceptions that animals, their lives and experiences are not be taken seriously. The debunking articles and the discursive power of the charge of anthropomorphism in this case reinforced the trivial status afforded to animal stories generally and reorganised the meanings of the images to leverage comedic value from the public willingness to assign emotional capacities to kangaroos.

What constitutes an authority in this context and how credibility was assigned and understood was clearly played out across the press discourse as it developed over three days. The shift from the privilege afforded to a first-hand witness account to that of a scientific account was loaded with gestures towards normative assumptions about science's rational, objective status which was endowed with the discursive power to close down the story of the three kangaroos. The story of the kangaroos ended up being a cautionary tale about non-scientists mistakenly sentimentalising nonhuman animals by attributing them with emotion. Summarising the problem in an article that bore the headline ' "Mourning" kangaroo was trying to mate, says expert', one news media outlet quoted the following scientific explanation: 'There is a strong bond between the mother and the young but it's hard to attribute emotions to those sorts of situations. [. . .] These are not little people, they are kangaroos' (BBC News, 2016).

In the initial reporting of the story the photographer was granted the status of eyewitness to the kangaroos' interactions, and the claim to being present at the event in time and space afforded the account a fundamental reliability. Evidentiary proof in the form of the photographic record then placed the viewer of the images in the position of witness to the mediated event. 'Being there' is fundamental to the status of an eyewitness, with the ability to communicate what happened through words or other means, trustworthiness and competence considered equally important traits (Mortensen, 2014). The debunking articles undermined the implied claims to competence and trustworthiness that upheld the witness account; the broader cultural acceptance of 'science' as a source of rational objective truth about animal behaviour used as a common-sense antidote to the sentimentalised public reading of the images. Science, as it was presented by the press discourse, assumed a conclusive claim to truth of interpretation via a narrative that unfolded over three days of press coverage. The later articles used the term 'expert' to refer to the scientists, citing their professional titles and thereby displacing the photographer as having a claim to any truthful interpretation of the events. References to kangaroo emotion and the affective dimensions of the initial articles were set in scare quotes – particularly the words 'heartbreaking', 'mourning' and 'grieving' – and the most frequently used quote from a scientist provided both the serious and popular press with a soundbite that was presented as evidentiary, an opportunity for a viewer to corroborate themselves, and was utilised to amplify the humour of the situation: 'The evidence is here sticking out from behind the scrotum'.

Over the three days of reporting, the press discourse altered the emotional appeals of the kangaroo images to modulate the affective dimensions of the story and shift genre references in such a way as to regulate the emotional identifications of the viewer through denial of an emotional lifeworld to the kangaroos. A different narrative was entirely possible if other experts had been consulted – those with expertise in the study of animal emotion for instance – or, if questions about the emotional lives of kangaroos had been asked. The point here is that the media discourse on the events constructed a narrative in which kangaroo emotion was erased. Instead, the narrative borrowed certain genre cues to frame the audiences' understanding of the images, a meshing of natural history with melodrama and comedy that utilised familiar gendered dynamics to reinforce the trivial status of the events. In the follow up articles, the discourse made visible the tensions between the two interpretations. The original gendered encoding of the tragedy of the events was primarily signified by the image that showed the outstretched arms of the dying or dead female kangaroo towards the young kangaroo while the scientific reading of the image refocused attention on the male kangaroo's penis. Descriptions of the male kangaroo as 'horny' and 'hot and bothered' constructed a humorous story of male sexual confusion, amplified the absurdity of kangaroo emotions, replacing it instead with a reductive stereotype of naturalised male sexual drives.

Despite the intervention of the scientist's statements, the semiotic remediation of the kangaroo images continued to rely on normative gendered assumptions about motherhood and male aggression. Such cultural representations of kangaroos are, as Peta Tait notes in her analysis of the contradictory status of the kangaroo, well established (Tait, 2013: 176). Barbara Crowther notes in her analysis of natural history programmes, gendered behaviour is a favoured theme of the genre where anthropomorphic language is used to describe it 'in terms that play on assumptions specific to the writer's culture' (Crowther, 1995: 128). Crowther's observations are applicable to this case, where she notes that the language used by biologists and scientific journalists to describe animal behaviour show markers of patriarchal concepts through the 'insidious drip-drip of metaphors around mothering and fathering roles' (Crowther, 1995: 128). Crowther goes on to note that there are three distinct narrative structures used extensively in wildlife programmes: 'the life-cycle story, the quest narrative and the triumph of science (culture) over nature – mastery over mystery' (Crowther, 1995: 128). The life-cycle story, she notes, focuses on a birth-to-parenthood rather than birth-to-death narrative while the language used to describe animal behaviour on screen more often confers active agency to the male. Both readings of the kangaroo images, as they were constructed within the press discourse, utilised the familiar gendered rhetoric. In doing so, both relied on anthropomorphic framing of the kangaroos' interactions – an inevitability of human-to-human communication – despite one of the readings being positioned as the scientific, objective discourse that claimed to eschew anthropomorphic interpretation.

In the case of the grieving kangaroos, the scientists' interpretation of the images was constructed by the press discourse as the definitive assessment of the event. The humour of the situation was validated by the gendered norms of sexualised comedy which obscured questions relating to the wellbeing of the individual kangaroos and overrode any serious consideration of their emotional lives. Some questions were raised online by those concerned with the welfare of the joey and others who queried if the female kangaroo had been carrying any other young in her pouch. It is not unusual for a female kangaroo to 'have a young joey hopping along beside her and still feeding from her milk, a baby in the pouch and a foetus in suspended animation in the womb' (Simons, 2013: 25). However, mainstream media accounts were primarily concerned with debunking articles on the 'problem' of anthropomorphism rather than asking questions about the welfare of the surviving joey, about what had happened to the body of the female kangaroo or about the type of emotional inner lives that kangaroos do have. There is an important issue with regard to the potential for moral concern to be shut down or extended in these situations. Arguably, the emphasis on constructing kangaroo emotion as erroneous sidelined or even erased questions of concern about the wellbeing of the surviving kangaroos. At the site where the kangaroo's experience was imaginatively interpreted as one of emotion, there was an opportunity for intersubjective cross-species identification that allowed for an expression of care for the wellbeing of those kangaroos who had survived.

In this case, it is not simply a situation in which the objectivity of science reasserted itself over popular sentimentalised subjective anthropomorphism. It certainly reveals the contest over the authority of interpretation and the discursive erasing of animal emotionality in a specific species. But we cannot suggest that this was a clearly won battle between non-anthropomorphic science and the anthropomorphic popular. Both discourses attributed intentionality and both appropriated gendered and genre cues to make sense of the situation. If social media shares are to be taken as a proxy for public engagement, then all metrics point towards the grieving kangaroo story being more popular than the debunking article. Even where the corrective article was shared, there was clear resistance on social media to the scientific explanation. That anthropomorphism was constructed as a denigrated object of discourse and that the emotional lives of kangaroos can be regulated and managed in this way demonstrates a certain ambiguity that is, I contend, species-specific. We have to take into account here the ambiguity of kangaroos who are constructed as humanlike within some aspects of popular culture (such as the television programme *Skippy* for a certain generation) but are also thought of as 'pests', as iconic national symbols, as food, and endorsed in some spheres as 'heart-healthy meat'. The ambiguity of their meaning is driven in large part by their economic worth in different domains therefore the contest over the authority of interpretation of the three kangaroos in Queensland is inescapably tied to the entanglements of capitalism and mediation that provide the shifting contexts for this site of anthropomorphism. What then of the emotional capacities of animals whose cultural meanings are less fluid, and how might this relate to anthropomorphism and questions of gender and genre?

Canine emotion

Emotion has been intrinsically linked to the female body, whereby a dominant discourse of gendered assumptions aligns aggression and rationality with masculinity and irrational emotion with women. These gendered associations have had social consequences. Although the dualism was not by any means absolute, the idea that women and emotions should occupy the private sphere while the public sphere was the rightful place of men and rationality has predominated. Such gendered relations have been reflected in popular culture where representations of emotional excess were a defining feature of film and television genres that sought to address female audiences through their focus on what were considered to be 'women's concerns'. Maintaining the public/private divide, domestic spaces were frequently the settings for representations of female emotional excess. Genre classification reveals one way in which emotion was feminised in the twentieth century with, for instance, 1930s melodrama referred to as 'the women's film' and later the soap opera being widely thought of as a 'women's genre' due to the high emotional content of both forms. The derogatory descriptors 'tear-jerkers' and 'weepies' referred to the climactic scenes of women's films that often depicted some type of loss or near loss, the intention of which was to bring audiences to tears.[1]

Although less well discussed than the women's film, tear-jerker films that feature nonhuman animals have long been a commercial Hollywood mainstay. Denigrated for their schmaltzy sentimentality, movies such as the *Lassie* features, *Old Yeller* (1957), *The Fox and the Hound* (1981), *Where the Red Fern Grows* (1974), *Homeward Bound: The Incredible Journey* (1993), *Marley and Me* (2008) and *A Dog's Purpose* (2017) have proved financially successful at the box office and in post-theatrical sales. Themes of loss, sacrifice, suffering and family variously inform the emotional appeals of such films as the characterisation of canine subjects draws liberally on the genre cues of melodrama. The focus on dog films here is purposeful. In the categorisations of top-grossing commercial films, those that feature dogs, horses, mice/rats, dinosaurs and dragons outnumber other nonhuman animals, real or mythical.[2] Of these, dog films outnumber the rest by some margin and of the top-grossing live action family films that feature non-talking nonhuman animals, five of the top ten have a dog as the main character. In the family live action talking nonhuman animal category, the situation is different, and only two of the ten top grossing films feature dogs. Instead, the talking live action list is dominated by nonhuman animals who are depicted as mainly bipedal (five of the ten). It seems that when it comes to emotional resonance, in commercial feature films at least, dogs have a greater affective capacity when they do not speak (but see Chapter 5 for my discussion on talking dogs).

The emotional lives of nonhuman animals are often trivialised, in part a legacy of the alignment of gendered norms with certain modes of anthropomorphism. Dogs and cats, in other words the most popular companion nonhuman animals, are those who feature most frequently in affectively engaging cultural forms such as fiction films (but also memes, viral video clips and so forth) *and* are depicted as emotional beings. Popular culture plays an important role in that it reinforces the idea that we feel greater empathy for those nonhuman animals with whom we are most familiar, and those species are more likely to be attributed with the capacity for emotion. In other words, popular culture perpetuates a discourse of reciprocal emotion that is often trivialised and commonly regarded as anthropomorphic. Thus, while it might normalise the emotional lives of nonhuman animals, at the same time that emotionality is managed through the mode of its presentation.

While the emotional lives of kangaroos were contested by a discourse of scientific rationality, narratives of grieving dogs enjoy a wider acceptance and acknowledgement. Popularised narratives of canine grief have preserved certain aspects of continuity across nineteenth-, twentieth- and twenty-first-century iterations – a demonstration of their long-standing appeal. In these cases, a dog is usually depicted as a loyal yet innocent, witness to human goodness or tragedy. Whilst science may reject some accounts of nonhuman animal emotion, stories of canine grief have retained a cultural legitimacy contextualised by the specific social arrangements of the time. Edwin Landseer's *The Old Shepherd's Chief Mourner* painted in 1837 depicts a black-and-tan sheepdog resting his head on a wooden coffin. The objects in the painting reference the occupation and morality of the dead man, although the main subject is not the death of a 'good shepherd' but rather the expression of canine grief. From a twenty-first-century perspective Landseer's painting is characteristic of an anthropomorphic nineteenth-century sensibility that was rejected by early twentieth-century modernism. Steve Baker (2000) writes about this denial of subjective interpretation and anthropomorphism within art and concludes, 'As the example of modernist art history as a whole suggests, the animal comes to be least visible in the discourses which regard themselves as the most serious' (Baker, 2000: 21). Baker argues that modern art had to remove all visible traces of the nonhuman animal to effectively banish the 'memories of the unashamedly anthropomorphic sentiment of an earlier age' (Baker, 2000: 20). If anthropomorphised animals had no place within serious art, they remained eminently visible within twentieth-century popular culture. In addition to *The Old Shepherd's Chief Mourner* (1837), Greyfriars Bobby (circa. 1858 and *Greyfriars Bobby* 1961), *Caesar the King's Dog* (1910), Hachi-Ko (circa. 1925), Lassie (*Challenge to Lassie* 1949), *Hachi: A Dog's Tale* (2009) and press accounts of 'Squeak' (2002), 'Hawkeye' (2011), 'Leao' (2011), 'Spot' (2012) and 'Ciccio' (2013) are all popularised narratives of canine grief.

Maud Earl's 1910 image of *Caesar* was produced following the death of Edward VII. There are obvious similarities between Earl's image and Landseer's painting. The dog occupies the central position within the image while an empty chair references the death of a human. In the case of Caesar however, the image refers to a real dog, the fox terrier companion to Edward VII. Photographic portraits of the King and his dog were widely circulated in postcard form and the dog's devotion and loyalty to the King was well-known to the public before Earl's painting. On the occasion of the King's funeral, Caesar had a place in the official procession where he followed the carriage containing the body and the attending heads of state were instructed to walk behind him. Newspapers reported that the dog had headed the procession, and three days after Edward's funeral the first edition of *Where's Master?* authored by Caesar The King's Dog, was published. The book was an immediate success, and between 13 June and 20 September 1910, 12 editions of the book were published and sold out. Maud Earl's image, *Caesar*, was included as an insert on the first pages of each edition of *Where's Master?* and the image was also reproduced in the *Illustrated London News*. By the Christmas of 1910, the best-selling toys were 'Dog Caesars', reproduced as plush stuffed dolls or plaster models and 'Dog Caesar' calendars had sold out. In the *Daily Mail*, a local toy merchant was quoted as saying,

> Who would have thought in the trade that people would have nothing but Dog Caesars? You see every mortal child that came in here wanted the dog. They would not look at Teddy bears, and we have to save them over for next year.
>
> (Grey, 1910: 4)

The ideal of the loyal, brave dog proved particularly potent. It suggested that dogs are autonomous individuals who were prepared to go above and beyond any training or instinct to serve human interests. Such qualities have currency in the wider public domain where 'loyalty', in the case of Caesar, was configured as an idealised relationship between subject and state. Caesar as the author of a book about his relationship to the King provided an authoritative witness account of 'his master's' good character that simultaneously expressed the expectations of loyalty from subjects to the monarchy and nation.

In 2002 *The Sun* newspaper reported on the death of farmer Terry Ford in Zimbabwe. The headline, which stated 'They killed my master', constructed the dog as witness. The report was accompanied by an image of a dog curled up next to the covered body of Terry Ford and described how Ford had been murdered during Robert Mugabe's land reclamation project (*Sun*, 19 March 2002: 24). The dog, Squeak, refused to leave the body until Ford's partner managed to coax the dog away. Between 19 and 23 March 2002, the *Sun* carried three reports on Terry Ford's death and funeral. Each article was accompanied by an image of the dog who also featured in each headline; 'They killed my master' (*Sun*, 19 March 2002: 24); 'Loyal terrier Squeak is safe' (*Sun*, 20 March 2002: 2) and 'Loyal to the last' (*Sun*, 23 March 2002: 20). In this case, the identity of the dead man was understood in relation to the loyalty and grief of the dog. Similarly, US news reports covered the story of a dog named Hawkeye who lay in front of Navy SEAL Jon Tomlinson's casket at his funeral in 2011. In Italy, the human companion of a dog named Ciccio was described as 'Maria of the fields', a regular churchgoer, and the human companions of two mourning dogs named Leao and Spot were victims of tragic circumstances. As these examples suggest, the normalisation of canine emotion within popular culture has persisted to serve differing ideological and human interests. In these narratives canine emotion is constructed as a means by which the character of humans can be realised, and although positioned as detrimental to the progress of science in the case of other species, the emotional capacities of canines generally has retained cultural legitimacy. Such narratives of canine loyalty reinforce a particular type of idealised human-nonhuman animal bond that can be leveraged in other spheres where the attribution of emotion to animals other than humans is carefully managed under the watch of corporate capitalism. The advertising of 'pet products', particularly dog foods, has continued to position humans as the guardians of canine health and emotional wellbeing, the two being regularly conflated in marketing messages. The emotional lives of dogs have been privileged in this sense and proved useful as moral exemplars as well as being economically productive. In other words, capitalism finds places where the emotional lives of other animals can be accommodated; where the expression of similitude proves economically advantageous. While advertising presses consumers to acknowledge the emotionality of dogs and purchase commodities to ensure their happiness and health, elsewhere the emotional lives of other sentient beings are denied, ridiculed or trivialised to ensure their continued exploitation.

Happy cows

There is no doubt that the rise of 'happy' farmed nonhuman animals in advertising and packaging imagery widens the repertoire of representations of 'emotional animals'. Yet there is a distinct difference between marketing imagery where the 'happy cow' is constructed as a one-dimensional emotional character compared with, for instance, films that feature dogs. While their characterisation remains relatively simple, dogs tend to be granted a wider emotional repertoire in fiction films to accommodate the various dramatic twists and turns that build the emotional address of such narratives. With a few exceptions, in popular culture generally, pigs, cows and chickens are not granted the same levels of emotional complexity and this plays well to forms of anthropocentric exploitation. By limiting representations of the emotional capacities of farmed animals to narrow depictions of happiness, their exploitation is much easier to manage. When the lives of pigs and chickens have been imagined by popular culture in films such as *Babe* and *Chicken Run*, commentators on both sides of the debate about the wrongs of carnism have claimed that the anthropomorphic depictions might persuade audiences away from eating meat. But, these filmic depictions compete in an image landscape that has been dominated by a trend in marketing nonhuman animal bodies as commodities and which also attribute nonhuman animals with emotion. From this tendency in advertising has emerged, for example, the imagery of 'suicide food' (Grossblatt, 2011) – depictions of nonhuman animals who act as though they are happy and willing to be consumed. With little to redress the balance in terms of dominant depictions, images of happy 'suicide food' act on a paucity of public information about the material realities of the lives of animals in industrial farming and food production systems. Such imagery plays with the idea that nonhuman animals are not only happily complicit in their own exploitation and deaths but that it is valid to treat their deaths with humour. Marketing companies have thus managed nonhuman animal emotionality into representational absurdity, often masquerading as 'cool' irony and pastiche that popularises and makes acceptable the notion that their slaughter can be trivialised and that humour is an appropriate lens through which to view the abuse of other sentient beings.

The imagery of cows within popular culture is primarily dominated by marketing industries which have broadly embraced anthropomorphism as a 'tool in order to either position new brands or reposition existing ones' (Pomering and Frostling-Henningsson, 2014: 150). Smiling cows on the packaging of dairy products function to reassure consumers of two interconnecting myths: that dairy cows are happy and that happy cows 'produce' a better quality 'product'. The associations with freshness, wholesomeness, countryside lifestyles, pastoral calm and a free roaming existence are far removed from the realities of milk production. Indeed, the image of a smiling cow is not only commonplace but it has also been central to the brand success of a number of dairy products as well as to the California Milk Advisory Board's advertising campaign: 'Great milk comes from Happy Cows. Happy Cows come from California'. Moreover, the notion of the happy farmed animal has been instrumental in realigning the animal-industrial complex with the notion of the caring farmer; the industrial scale of nonhuman animal suffering can thus sit, albeit precariously, behind compelling imagery and narratives of family farms, traditional practices, welfare standards and the corollary of bovine emotional contentment. One 2012 op-ed about a family farm in the *New York Times*, under the headline 'Where cows are happy and food healthy', asked, 'Is it soggy sentimentality for farmers to want their cows to be happy?' only to reassure readers that emotional wellbeing equates to capitalistic gain: 'For productivity, it's important to have happy cows. [. . .] If a cow is at her maximum health and maximum contentedness, she's profitable' (Kristof, 2012). Images of 'downer cows' with foot infections, milk fever (hypocalcaemia), ketosis, or nerve paralysis; culling practices; concentrated animal feeding operations (CAFOs); cows forcibly inseminated and then separated from their calves and the plight of those calves after their birth – in other words the realities of industrial farming – seldom make their way into the public domain. It is instead the mainstream media stories about family farms and the images of a cartoon cow with the 'winning smile' and a daisy in her mouth or a Holstein or Jersey breed gently grazing in the bucolic landscape which reassure consumers that welfare standards are writ large across the joyful face or contented bovine body and that their 'milk comes from a good place' (American Dairy Association North East, 2017). As Matthew Cole argues in relation to the concept of 'happy meat',

> in conceding sentience and an expressive self, while continuing to confine and kill for gustatory pleasure, 'happy meat' and 'animal friendly' welfare discourses attempt to remoralize the exploitation of 'farmed' animals in such a way as to permit business as usual, with the added 'value' of ethical self-satisfaction for the consumer of 'happy meat'.
>
> (Cole, 2011)

In much the same way as it does with 'happy meat', the animal-industrial complex hijacks the emotional lives of cows and other farmed animals and distorts them through various iterations of 'happy' cow imagery.

These strategies, while effective, are not immune to the correctives to public knowledge that have been advanced by, for example, the Go Vegan World campaign and the film *Cowspiracy*. In the former case, a decision by the Advertising Standards Agency (ASA) – the UK's independent advertising regulator – found in favour of the campaign following complaints that 'the ad did not accurately describe the way that dairy cattle were generally treated in the UK' (ASA, 2017).

The advert appeared in the UK national press in February 2017. The headline 'Humane milk is a myth. Don't buy it' appeared above an image of a cow whose gaze directly addresses the viewer. She stands behind barbed wire and against a black background. Below the headline, and above her head, smaller text stated:

> The mothers, still bloody from birth, searched and called frantically for their babies. Their daughters, fresh from their mothers' wombs but separated from them, trembled and cried piteously, drinking milk from rubber teats on the wall instead of their mothers' nurturing bodies.
>
> (Go Vegan World advertisement, 2017)

The complaints, some from the dairy industry, suggested that the statements were misleading and the campaign group were required to substantiate the claims. Go Vegan World provided evidence that supported the phrases 'still bloody from birth' and 'fresh from their mothers' wombs'. The group made the point that the ad did not imply that separation was occurring prior to the 12 to 24 hours recommended by the Department for Environment, Food and Rural Affairs (DEFRA). Instead, the point was to comment on the injustice of the separation. In its final assessment, the ASA did not uphold the complaints and finding in favour of Go Vegan World stated:

> Although the language used to express the claims was emotional and hard-hitting, we understood it was the case that calves were generally separated from their mothers very soon after birth, and we therefore concluded that the ad was unlikely to materially mislead readers.
>
> (ASA, 2017)

The ASA ruling was covered by all the major UK news outlets, the majority of which used content from the Associated Press that differed little from the wording of the ruling from the ASA. Headlines gave some indication of the position of individual news outlets, most of which referred to the advert as 'vegan' with some making reference to a battle between vegans and farmers or the dairy industry: '"Humane milk is a myth" ad relaunched after ASA rejects farmers' complaints' (Bird, 2017); 'the dairy industry tried to block this vegan advert form appearing' (Mills, 2017). What was significant about the advert was that it described the realities of cows' and calves' experience as sentient beings with emotional lives. The mode of address was in the form of a witness account, a point of contention for those who challenged the legitimacy of the claims. Inevitably the subjectivity of the style of discourse was contrasted with the notion of scientific objectivity in one report which stated that the advertisement used 'highly emotive and anthropomorphic language, applying human emotions to a calf or dairy cow' (Webster and Lloyd, 2017). The statement was credited to the head of public affairs at the Royal Society for the Prevention of Cruelty to Animals (RSPCA), who stated, 'I do not believe that we have scientific evidence to support that' (ibid.). Such claims reveal the precarious nature of the industry's discourse on dairy that had, in the previous month, promoted World Milk Day using the hashtag #HappyCows across social media and which was, subsequent to the ASA ruling, used by farmers to post images and videos of cows. The discourse continued to conflate nonhuman animal happiness with welfare standards, a stance that relies on the denial of cow grief and sadness while promulgating the notion that the human interests that inform welfare standards will guarantee cow happiness. Cows are thus selectively attributed with happiness by the industry discourse; a form of emotional editing that confers to nonhuman animals only the qualities that are intended to reassure consumers of welfare standards, irrespective of how inhumane such standards are in the first place. In this context, the careful management of nonhuman animal emotion in service to capital walks a particularly tricky line and is open to challenge. Taking into account that the ASA found that the Go Vegan World would not materially mislead the public, situated empathetic appeals that acknowledge degrees of similitude between humans and other animals based on an informed understanding of that nonhuman animal's reality (rather than the human interests that inform welfare standards) can be highly effective. Whether or not those appeals are sites of imaginative or commodified anthropomorphism is a matter of context and we can set the example of cows against that of dogs and kangaroos to reveal how individual species are situated in relation to claims to truth that benefit capitalistic enterprise.

Happiness and pleasure: misunderstandings and interventions

The popular understanding of nonhuman animal happiness generally has tended to rely heavily on highly anthropocentric views of emotional expression. Jonathan Balcombe argues in relation to fish, for example, that 'because [they] don't have expressive faces, we tend to find them difficult to identify with or feel for' (Balcombe, 2016: 148). On the other hand, dolphins have fared quite differently, and while Balcombe rightly points out that this could be partially due to the popular understanding that dolphins have large brains, dolphins also have a fixed expression (a curved shape to the mouth) that humans read as happy or smiling. Not all dolphin species have the trait, but the bottlenose dolphin – the one most commonly seen in film, television and theme parks – has the particular anatomical configuration that gives the illusion they are smiling. Alan Bauch notes that 'the smile is permanent and masks the mood of even the angriest dolphin, but', he adds, 'overall it is a trait that endears the animals to us even further' (Bauch, 2014: 10). This popular misattribution of happiness to dolphins has resulted in unacknowledged widespread suffering, a theme that is taken up in in the film *The Cove*, when dolphin advocate and the former trainer of dolphins for the television series *Flipper*, Richard O'Barry claims, 'A dolphin's smile is the greatest deception. It creates the illusion that they're always happy'. O'Barry also used the popular misconception as the basis for the title of his book *Behind the Dolphin Smile*, in which he recounts various instances of dolphin suffering which have occurred while humans believe that they are happy or laughing (O Barry, 2012:

5–7). Although they have been widely portrayed as happy, fun-loving beings, the suffering of many dolphins is, in part, a consequence of our imagined emotional similitude whilst the suffering of fish, for example, has been ignored due to the value we place on facial expression.

The eco-documentaries *Blackfish* (2013) and *The Cove* (2009) have directly challenged myths of cetacean happiness. Part of a cluster of environmentally themed documentaries that has grown since the success of *An Inconvenient Truth* in 2006, the eco-documentary is an issue-driven film. In broad terms, eco-documentaries explore a wide range of issues that tend to highlight the impact of capitalist industries and corporate control on animals and the environment. Mainstream interest in environmental issues combined with a resurgence in the popularity of feature-length documentaries and increasing opportunities for their distribution have provided the context for the emergence of eco-documentaries. Although not the primary theme, *Blackfish* and *The Cove* do counter popular ideas about dolphin happiness while at the same time constructing the nonhuman animal subjects as emotionally complex beings. Crucially, these and other eco-documentaries have been successful in engaging audiences and mobilising action, the emotional appeals of the films being a key element of that success (see Chapter 6 for an extended discussion of *Blackfish*). Similarly, a social media campaign by International Animal Rescue (IAR) intervened in the public discourse on keeping slow lorises as pets and was effective in remediating footage of an individual slow loris to reorganise the empathetic lens through which she was viewed. In this case, the emotional appeals of a video made as part of the campaign and circulated widely on social media platforms reframed what had previously been regarded as 'cute happy' footage as a depiction of a slow loris suffering. The discourse highlighted how human-nonhuman animal difference can lead to a misunderstanding of species-specific suffering while also making empathetic appeals on the basis of human-nonhuman animal similitude in terms of emotional experience. In this case, however the campaign did not address the practices of a corporation or commercial industry but public misunderstanding and an illegal pet trade. In this situation, the claims to truth made using anthropomorphic strategies raised no discernible opposition and as such the contest over authenticity of representation and interpretation found little in the way of challenge.

In 2015 the 'Tickling Is Torture' campaign launched by IAR aimed to highlight the suffering of slow lorises, victims of a pet trade that the welfare organisation cited as 'the biggest threat to the survival of the species' in Indonesia (IAR, 2015a). A trend for keeping slow lorises as companion animals had, according to IAR, intensified significantly since 2009 due in large part to the circulation of viral videos across social media that depicted the primates being 'tickled', holding umbrellas and eating rice balls. The IAR campaign stressed the widespread mistaken interpretations of the animals' behaviours in these videos, the most popular of which featured a slow loris named Sonya who was shown leaning against a green patterned duvet with her arms held straight above her head while human hands 'tickle' her. Around one minute in length and shared under various titles that included 'Slow Loris Tickle: Sonya the tickle lover', 'Slow loris loves getting tickled' and 'And now . . . at last – Sonya!!!! (slow loris)', one version of the video ('tickling slow loris') alone received in excess of nine million views on YouTube before being removed in 2012 (Nekaris et al., 2013). While slow loris videos regularly featured in top viral video lists usually accompanied by some reference to the primates' 'cute' appearance and the pleasure they derived from their interactions with humans, the IAR campaign explained that tickling a slow loris was a form of torture. By lifting her arms Sonya was not signalling enjoyment or happiness as many viewers supposed. Instead the slow loris, the only known venomous primate, was recruiting venom from brachial glands in her arms which would usually be accessed through grooming, stored in a tooth comb in her mouth and transferred through a bite when threatened. When being 'tickled' Sonya, far from finding the situation pleasurable was, IAR argued, terrified (IAR, 2015b).

As part of the campaign International Animal Rescue produced a three-minute video that featured the actor, Peter Egan, to act as a corrective re-reading of the 'cute' slow loris videos. Slow, melancholic piano music plays throughout the video, the melodic shape, tonal vocabulary and instrumentation reinforcing the rhetorical appeals of Egan's commentary on the plight of the slow loris. Following three clips from slow loris viral videos, Egan's piece to camera begins:

> You may have seen YouTube videos of a cute little animal called a slow loris being kept as a pet and tickled. But would you think it was cute if you knew that tickling a slow loris is actually torturing it? And, that slow lorises are suffering terribly as a result of these videos.
>
> (IAR, 2015b)

Egan's slow, measured direct address describes the practices routinely employed within the exotic pet trade and poses a question at the end of each factual statement: 'Would you still want to watch it and share it with your friends?'; 'Would you still think it's cute?'; 'Now, do you still want to support this trade?' Video footage of slow lorises having their teeth removed without anaesthetic and still images of live and dead lorises crammed together in small cages accompanied Egan's commentary.

The press response to the campaign was key to driving up the number of views that the video received and the subsequent shift in public attitudes. The change in public comments on slow loris videos on YouTube, particularly the video of Sonya being 'tickled', was apparent within weeks after the release of the 'Tickling is Torture' video. The link to the IAR video and the campaign website was frequently posted in comments beneath the videos which were still receiving high volumes of views. In this way, the comments sections for the 'tickle' videos were also active in disseminating the IAR 'torture' message. The public conversation about slow lorises changed with comments no longer saying 'cute' but instead referring to the videos as 'torture', 'abuse' and 'cruel' and accompanied by demands that the videos be removed from social media platforms. Comments that also referred to the loris's recruitment of poison in her elbows and to the practice of tooth removal demonstrated that the IAR message was circulating successfully.

In a random sample of comments made in a one-month period after the IAR video was distributed in relation to a re-shared video of 'Sonya', 96 per cent were negative, with 23 per cent of these making explicit reference to the poison recruitment and tooth extraction, a further 34 per cent using the terms 'abuse', 'torture' or 'cruel' and 11 per cent using another term to describe the loris's emotional state. Other comments called for the video to be removed or made reference to the loris being wild and/or endangered.[3] The YouTube channel 'SlowLorisChannel', with 48,000 subscribers, disabled public comments on its 11 loris videos.[4]

It is useful to compare the slow loris case to that of the grieving kangaroos, the happy dogs and cows. The meanings of the loris as a species are that of 'cute', vulnerable and free roaming. The pet trade that has grown up around the slow loris is illegal and there are less immediate benefits to western capitalism in managing the emotional life of the loris than we find in the case of dogs whose emotional life has been leveraged for commercial benefit and cows where regulation of their emotional states through marketing serves industrial interests while distancing the consumer from the realities of their lives within the animal agriculture system. The ambiguity of the kangaroo, her meanings having to accommodate being iconic symbol, pest and food make the contest over the authority of the interpretation of her life and emotional experiences messier than that of the loris, who has to bear a less diverse set of meanings. The genre and gendered cues that give context to these sites of anthropomorphism are similar but the ways in which they are organised to elicit imaginative and empathetic connections differs considerably.

Conclusion: the currency of animal emotion

The desire for a mirror of our own emotional expression extends to other species. In the entertainment industries chimpanzees have been trained to 'smile', to bare their teeth in what approximates a human smile but is instead actually a fear grimace – usually reserved for situations when the individual is fearful or anxious. In horses and lambs, the lip curled back to expose the front teeth, sometimes referred to as the flehmen response, can indicate pain or is more commonly associated with odours or tastes of particular interest. Horses are trained to produce the flehmen response on command – an approximation of the human 'smile' or 'laugh'. In the case of the television show *Mister Ed*, a horse named Bamboo Harvester was trained to perform the behaviour on command to suggest that the eponymous star was 'talking'. Exploitation of the horse, cow, chimpanzee and dolphin 'smile' within popular culture has been prevalent and reflects the casual narcissistic anthropocentrism at sites of commodified anthropomorphism.

During the twentieth century, representations of the emotional capacities of animals were widely excluded from serious and scientific discourses but they predominated within popular culture. The impact of behaviourism during the first half of the twentieth century that designated anthropomorphism as methodologically unscientific cemented such divisions and aligned the humanisation of animals with fiction rather than 'science fact'. At the same time anthropomorphism was being vilified in some spheres, cultural narratives of emotion have referenced both 'fictional' and 'real' animals, differentiating the latter stories from the less ambiguous anthropomorphic depictions of, for example, talking animals. In this sense, the attribution and editing of nonhuman animal emotion can seem plausibly 'real' when set against a cultural anthropomorphic spectrum that includes talking, singing, and dancing pigs and mice. While knowledge conditions have at times militated against the scientific authorisation of anthropomorphism, the attribution of emotion to animals has retained a level of popular cultural credibility.

We remain fascinated by happiness, particularly our own. In 2011, the UN General Assembly invited member countries to measure the happiness of human populations as a means to inform policy-making decision. The resulting *World Happiness Report* endeavoured to link happiness to economic growth, a neoliberal quantification of emotion so that it bears easy translation into gross domestic product (GDP). Tracing the cultural history of the emoji, or digital pictogram, Luke Stark and Kate Crawford (2015) note that it has its beginnings in the 'iconic "smiley" face' of the mid-twentieth century. 'The original smiley', they contend, 'crystallized the force of the feeling into an icon that could be simultaneously mobilized in the service of institutional corporate power, transcend that control to become a cultural touchstone, and become recaptured as a form of intellectual property' (ibid.). The smile, they argue has become 'an artefact of Capital'. Yet, nonhuman animal smiles have been in the service of capital for much longer.

Happy dog imagery has flooded the internet, often presented as an ironic counterpoint to grumpy cats, although the latter has by far the greater presence and cultural weight on social media. Such is our obsession with the smile and the authenticity of expression that various online tests are available to measure users' ability to differentiate between 'real' and 'fake' smiles. The payoff for getting the right answer is that those who can identify the 'real' smiles are considered to be endowed with greater levels of empathy, a capacity that could bear being extended to our understanding of other-than-human expressions of happiness. What is crucial, however, is that the notion of emotional authenticity has been foregrounded and there is a continuing public fascination with how emotion is performed. No doubt there is a general cynicism over the outward display of human emotion and the apparent ease with which it can be fabricated. In the case of nonhuman animals, the attribution of emotional lives and the perception of authentic emotion are no less complex and contested.

In the teasing out of which nonhuman animals qualify for attribution of an emotional inner life, the question of authenticity is continually entwined with the question of anthropomorphism and the currency of animal emotion remains valuable to capitalism. This currency is species-specific and therefore the anthropomorphism aligned with it is contextual and situated. Scientific discourse does not hold an absolute authority over the meanings attached to species nor to individual nonhuman animals. The mediation of kangaroo life and experience contextualised by scientific discourse may act to disrupt empathetic anthropomorphism and attempt to reconstruct distance between human and nonhuman animals but such forms of regulation and management are unstable. The case of the three kangaroos suggests that an understanding of familial relationships, the cross-species intersubjective identification with motherhood and the relationship between mother and child, which relied on genre and gendered cues can be understood as a site of imaginative empathetic anthropomorphism that resulted in the extension of moral concern for the surviving kangaroo. The currency of animal emotion however exists in an economic framework that has actively encouraged high levels of human consumer investment in the emotional labour of animals. The next chapter focuses on talking animals to take up the issue of the emotional labour and its reliance on the aesthetics of cuteness.

Notes

1 Although it might seem counter-intuitive to suggest that there is pleasure in sadness, Flo Leibowitz, argues that 'many tear-jerking scenes evoke sorrow and admiration at the same time, and part of the enjoyment of women's films is in the experience of these mixed emotions' (Leibowitz, 1996: 222).
2 At time of writing, the number of top-grossing commercial fictional dog films since 1974 is 63. Horse films number 32 since 1980. All figures taken from Box Office Guru database.
3 Random sample of 100 public comments over a one-month period at www.youtube.com/watch?v=PZ5ACLVjYwM.
4 Despite the apparent success of the IAR public education message, reports in 2016 suggested that the number of lorises being sold through the illegal pet trade was increasing and the primary sales platform was Facebook (Molloy, 2016 www.telegraph.co.uk/news/2016/10/21/suffering-slow-lorises-with-teeth-ripped-out-being-sold-on-faceb/).

References

Allan, S. and Thorsen, E. (eds) (2009) *Citizen Journalism: Global Perspectives*, Peter Lang, New York.
ASA (2017) 'ASA ruling on Eden farm animal sanctuary t/a go vegan world' 26 July, online at www.asa.org.uk/rulings/eden-farmed-animal-sanctuary-a17-381845.html
Baker, S. (2000) *The Postmodern Animal*, Reaktion Books, London.
Balcombe, J. (2016) *What a Fish Knows*, Scientific American/Farrar, Straus and Giroux, New York.
Bauch, A. (2014) *Dolphin*, Reaktion Books, London.
BBC News (2016) '"Mourning" kangaroo was trying to mate, says expert' 14 January, online at www.bbc.co.uk/news/world-australia-35308161
Bird, S. (2017) 'Humane milk is a myth ad relaunched after ASA rejects farmers complaints' *The Telegraph*, 26 July, online at www.telegraph.co.uk/news/2017/07/25/humane-milk-myth-ad-relaunched-asa-rejects-farmers-complaints/
CBS News (2016) 'Humans apparently mistook kangaroo necrophilia for tenderness' *CBSNews* website 14 January, online at www.cbsnews.com/news/kangaroo-necrophilia-mistaken-tender-moment-australia-daily-mail/
Cole, M. (2011) 'From "animal machines" to "happy meat"? Foucault's ideas of disciplinary and pastoral power applied to "animal-centred" welfare discourse' in *Animals*, Vol. 1 (1), pp. 83–101.
Crowther, B. (1995) 'Towards a feminist critique of television natural history programs' in Florence, P. and Reynolds, D. (eds) *Feminist Subjects, Multimedia*, Manchester University Press, Manchester and New York.

Formosa, A. (2016) 'Photographer captures kangaroo family's grief' *Fraser Coast Chronicle*, 13 January, online at www.frasercoastchronicle.com.au/news/his-heart-skipped-a-beat/2897351/

Fruen, L. (2016) 'I'll love roo till the end: Dying kangaroo mum reaches out to her joey in final moments' in *The Sun*, 14 January, online at www.thesun.co.uk/archives/news/123286/ill-love-roo-till-the-end-dying-kangaroo-mum-reaches-out-to-her-joey-in-final-moments/

Gage, A. (2016) 'Images of "Grieving" kangaroos misinterpreted' *Australia Museum* website, 14 January, online at https://australianmuseum.net.au/blogpost/museullaneous/images-of-grieving-kangaroos-misinterpreted

Grey, W.E. (1910) '"Caesar": The dog who belonged to a king' in *The Daily Mail*, 22 December, p. 4.

Grossblatt, B. (2011) 'Suicide food blog' online at http://suicidefood.blogspot.co.uk/

Horton, H. (2016) 'Heartbreaking pictures show mother kangaroo reaching for joey one last time before dying in male companion's arms' *The Telegraph*, 13 January, online at www.theguardian.com/world/2016/jan/13/photographer-mother-kangaroo-dying-queensland-australia

Hunt, E. (2016) 'Kangaroo in "grieving" photos may have killed while trying to mate, scientist says' *The Guardian*, 14 January, online at www.theguardian.com/science/2016/jan/14/photos-grieving-kangaroo-viral-but-scientist-says-sexually-aroused

International Animal Rescue (2015a) 'Tickling is Torture' online at www.international animalrescue.org/truth-behind-slow-loris-pet-trade?utm_source=CP&utm_medium=website&utm_campaign=tickling_is_torture&utm_content=chtfom_imagebutton

International Animal Rescue (2015b) 'The truth behind the slow loris pet trade and "cute" tickling slow loris videos', video online at www.youtube.com/watch?v=otTNxR8C4uE

Khomami, N. (2016) 'Photographer captures moment kangaroo cradles dying companion as joey looks on' *The Guardian*, 13 January, online at www.telegraph.co.uk/science/2016/03/15/heartbreaking-pictures-show-mother-kangaroo-reaching-for-joey-on/

Kristof, N. (2012) 'Where cows are happy and food is healthy' *New York Times*, 8 September, online at www.nytimes.com/2012/09/09/opinion/sunday/kristof-where-cows-are-happy-and-food-is-healthy.html

Leibowitz, F. (1996) 'Apt feelings, or why "Women's Films" aren't trivial' in Bordwell, D. and Carroll, N. (eds) *Post-Theory: Reconstructing Fim Studies*, University of Wisconsin Press, Madison, pp. 219–229.

Mills, J. (2016) 'The dairy industry tried to block this vegan advert from appearing' *Metro*, 26 July, online at http://metro.co.uk/2017/07/26/the-dairy-industry-tried-to-block-this-vegan-advert-from-appearing-6808843/

Molloy, M. (2016) 'Suffering slow lorises with teeth ripped out being "sold on Facebook"', 21 October, online at www.telegraph.co.uk/news/2016/10/21/suffering-slow-lorises-with-teeth-ripped-out-being-sold-on-faceb/

Mortensen, M. (2014) *Journalism and Eyewitness Images: Digital Media, Participation and Conflict*, Routledge, London and New York.

National Geographic (2016) 'The disturbing truth behind those viral kangaroo photos', 14 January, online at www.nationalgeographic.com.au/animals/the-disturbing-truth-behind-those-viral-kangaroo-photos.aspx

Nekaris, A-I., Campbell, N., Coggins, T.G., Rode, E.J., and Nijman, V. (2013) 'Tickled to Death: Analysing public perceptions of "Cute" videos on threatened species (Slow

Lorises – *Nycticebus* spp.) on Web 2.0 sites' in *PLoS One*, 24 July, online at http://journals. plos.org/plosone/article?id=10.1371/journal.pone.0069215

New York Post (2016) 'Grieving kangaroo may have actually been a horny killer', 14 January, online at http://nypost.com/2016/01/14/grieving-kangaroo-may-have-actually-been-a-horny-killer/

O'Barry, R. (2012) *Behind the Dolphin Smile*, Earth Aware, California.

Peters, D. (2016) 'Please don't die, mum: The heart-wrenching moment a mother kangaroo reaches for her joey one last time- before dying in the arms of her male companion' *Daily Mail*, online at www.dailymail.co.uk/news/article-3396905/The-heart-wrenching-moment-mother-kangaroo-reaches-joey-one-time-dying-arms-male-companion.html

Pomering, A. and Frostling-Henningsson, M. (2014) 'Anthropomorphism brand presenters: The appeal of frank the sheep' in Brown, S. and Ponsonby-McCabe, S. (eds) *Brand Mascots: And other Marketing Animals*, Routledge, London and New York.

Readhead, H. (2016) 'No, this kangaroo wasn't "grieving" – It was raping a dead female, experts say' *Metro*, 14 January, online at http://metro.co.uk/2016/01/14/no-this-kangaroo-wasnt-grieving-it-was-raping-a-dead-female-experts-say-5623631/

Simons, J. (2013) *Kangaroo*, Reaktion Books, London.

Stark, L. and Crawford, K. (2015) 'The conservatism of emoji: Work, affect, communication' *Social Media and Society*, Vol. 1 (2), online at https://doi.org/10.1177/2056305115604853.

Switzer, E. (2016) 'Facebook comment' online at www.facebook.com/photo.php?fbid=10 208215607539001&set=o.106556526064540&type=1&theater

Tait, P. (2013) 'Caught: Sentimental, decorative kangaroo identities in popular culture' in Boyde, M. (ed) *Captured: The Animal within Culture*, Palgrave Macmillan, Basingstoke.

The Sun (2002) 'Loyal terrier Squeak is safe' in *The Sun*, 20 March, p. 2.

The Sun (2002) 'Loyal to the last' in *The Sun*, 23 March, p. 20.

The Sun (2002) 'They killed my master' in *The Sun*, 19 March, p. 24.

Webster, B. and Lloyd, B. (2017) 'Milk can be branded inhumane, advertising chiefs tell farmers after vegan campaign' *The Times*, 26 July, online at www.thetimes.co.uk/article/milk-can-be-branded-inhumane-advertising-chiefs-tell-farmers-after-vegan-campaign-jwgblxmb7

5 When animals speak

Introduction

In the previous chapter I examined the currency of animal emotion. In this chapter I consider how sites of anthropomorphism utilise the emotional labour of animals. As humans, we have a particular cultural fascination with interspecies communication and therefore the sites where such interactions are imagined are important to explore. The focus here is on talking animals and, in this chapter, I look at some contact points between science and popular culture, where popular culture has appropriated scientific studies of animal language to construct fictional narratives of interspecies communication, and sites where intersubjective cross-species connections meet cute aesthetics. When it comes to thinking about talking animals the issue of 'cuteness' is foregrounded and has tended to be critiqued as a means by which animals other than humans are infantalised and trivialised. My aim here is to consider what can be recovered from anthropomorphism where it falls into the chasm bordered on one side by the sentimental mode and on the other by cute aesthetics – two aspects that dominate popular culture in an era of neoliberalism.

Animal language

In October 2012, the story of NOC (pronounced 'no-see'), a male beluga whale, drew international media interest. NOC, who had been captured in 1977 and lived in captivity at the National Marine Mammal Foundation, had spoken. What caught the public attention was a recording of NOC's vocalisations that had been made in the 1980s and which accompanied the online publication of an article entitled 'Spontaneous human speech mimicry by a cetacean' in *Current Biology* (Ridgway et al., 2012). Beluga whales are well known for their vocalisations – dubbed by some the 'canaries of the sea' due to the range of their vocal calls – and there were anecdotal accounts of beluga whales mimicking human speech, but this was the first time such sounds had been recorded. The article's authors pointed out that NOC's utterances were initially spontaneous although he was later trained to produce the vocalisations on command so that scientists at the facility could insert a rapid response pressure catheter into his nasal cavities and air sacs to analyse how the sounds were made.

The article gave a brief account of the scientists' first becoming aware of NOC's vocalisations:

> The whale was recognized as the source of the speech-like sounds when a diver surfaced outside the whale's enclosure and asked 'Who told me to get out?' Our observations led us to conclude the 'out' which was repeated several times came from NOC.
>
> (ibid.)

Several news outlets focused on NOC's agency and the motivations for his attempts at interspecies communication. One headline in the UK popular press declared that NOC had 'used an underwater microphone to make contact with scientists' (Prigg, 2012) while another claimed that he had 'made extraordinary efforts to make contact with his human captors' (Connor, 2012). Elsewhere, NOC's attempts to communicate were reported by the BBC as mimicry, the headline used scare quotes to emphasise that his utterances were merely 'human-like sounds' (BBC, 2012). *Nature*, much the same as the BBC and indeed the original *Current Biology* article, took the line that this was an example of mimicry, the point of interest for the report being how NOC made the sounds and not what he had said. *Nature* headed their article with 'The whale that talked' and used the subheading 'Captive beluga was able to mimic speech (sort of)' as a self-reflexive gesture to the irony of the headline. *National Geographic* also made use of scare quotes in its headline: ' "Talking" Whale Could Imitate Human Voice' (Scale, 2012) – a discursive act of distancing sarcasm. At the end of the *Current Biology* article, the authors cautioned 'reports of animal mimicry based solely on hearing vocalisations must be viewed skeptically' (Ridgway et al., 2012). To ensure that NOC's utterances did not trouble the safely guarded boundary of human exceptionalism, the *Huffington Post* quoted the lead author saying 'people should not think that whales can communicate on a conversational level based on these results' (Choi, 2012), and a marine biologist commented in the *National Geographic* article that it was unlikely that NOC understood the sounds he made.

What should we make of NOC's utterances? Some noted in online public forums that the recording sounded more like someone playing a kazoo than human speech and suggested that the desire for interspecies communication can deceive us to the extent that we hear language where there is merely sound. Was NOC 'simply' mimicking the human sounds he heard, was he ordering the diver out of the water or was he asking for the same opportunity as the diver, to be let out and away from his captive environment? Perhaps NOC was trying to communicate something to humans that, due to our paucity of understanding about beluga experience, we cannot possibly begin to comprehend. Indeed, this raises other questions about whether the issue is not with what NOC was or was not saying but if humans were able or even prepared to listen to him. And when we do listen, who should be granted the authority to translate what is being said? Margo DeMello summarises the difficulties when she writes 'Humans, from a position of superiority, can either choose to ignore what animals are saying, making them silent, or can interpret for them, which runs the risk of doing so from the human point of view' (DeMello, 2013: 5). The cultural mediation of animal experience will always be anthropomorphic to some degree, a human interpretation of the nonhuman animal mind, and therefore the risk, must surely be worth taking if the other option is silence.

Ludwig Wittgenstein famously wrote, 'If a lion could talk, we could not understand him' (2001: 190). Yet, as Erica Fudge points out, 'That's the danger. If we could hear them speak, we might not want to hear what they say' (Fudge, 2002: 127). According to scientists from a range of disciplines, there is little chance that humans will be faced with this particular dilemma as language remains a peculiarly human capacity. Features such as displacement, arbitrariness, productivity, reflexivity and cultural transmission have been ring-fenced together to uphold the view that human language is not only unique but that it confers species-specific privilege.[1] Although the idea of animal communication is not disputed, it is instead a distinction between communication and language that is purported to place humans in an exclusive position. Animals communicate, so the argument goes, but only humans have language, a specialist subset of communication the distinguishing features of which are rarely found together in any other nonhuman communication. For linguists Noam Chomsky and Robert C. Berwick, for example, language is species-specific and in evolutionary and genetic terms 'unique to the human lineage' (Berwick and Chomsky, 2016: 53). Language, they argue, evolved as a tool, not for communication, but for inner thought and only the human brain is 'wired up' in the right way for this (2016: 163–166). Espousing one of the general tenets of the discontinuity argument which, in evolutionary terms proposes that language emerged suddenly rather that in stages, Chomsky and others view it as a tool for thinking. And Chomsky and Berwick are certainly not alone in arguments that connect language and mind. So engrained is the link that the lack of a shared language between human and nonhuman animals is used to support arguments that discount the subjective experiences of other species and refute claims to nonhuman animal consciousness.

Within the tangled relationship between nonhuman animal communication and consciousness lies the issue of anthropomorphism, usually invoked alongside concerns that the attribution of language is misapplied anywhere other than within the human realm. For example, the work of primatologist Sue Savage-Rumbaugh has been met with criticism when she writes about bonobos in the following way:

> At times as I watch them, I seem to be staring into my distant past and seeing in front of me 'quasi-persons' – not people, but 'near people'. [. . .] With bonobos I experience a similar two-way understanding. I know how they feel, and they know how I feel.
>
> (Savage-Rumbaugh et al., 2001: 4)

Her ability to understand their feelings is possible, she explains, 'because of the expressions that emanate from their faces, the way they interpret the feelings of others, the depth of their commitment to one another, and the understanding of

one another that they share' (ibid.). Savage-Rumbaugh concedes that her interpretation goes against the norms of scientific observations but notes, 'This is a
perception I cannot shake off or dissuade myself from, no matter how often I try to
tell myself that I have no definitive scientific basis for these impressions' (ibid.).

Donna Haraway points out that other primates are particularly ambiguous
because they exist at the boundary of the 'almost human' (Haraway, 1989: 2).
External bodily similarity between humans and other primates has certainly motivated much of the research into animal language acquisition to focus on chimpanzees, gorillas and bonobos. Yet, as researchers during the 1960s discovered,
the vocal apparatus of apes differs substantially from that of humans in the ori-
entation of the vocal-laryngeal tract. Biological difference between humans and
other primates in the make-up of the vocal apparatus led researchers to experiment with other forms of communication, including the lexigram system favoured
by Savage-Rumbaugh and American Sign Language (ASL). As a consequence of
the appropriation of alternative forms of communication, the relationship between
language and 'talking' has been central to some critical appraisals of ape language
research findings. In response to Savage-Rumbaugh, linguist Talbot J. Taylor asks,

> from a scientific perspective, does Savage-Rumbaugh give overhasty, loose,
> even subjectively biased characterizations of Kanzi's response to hearing
> those utterances? Do Savage-Rumbaugh's characterizations make up a true
> and objective representation of what had occurred? Is she in fact justified in
> characterizing the scenes as she does?
>
> (Taylor, in Savage-Rumbaugh et al., 2001: 139)

Savage-Rumbaugh and Taylor's comments open up important questions about
who talks on behalf of other animals. DeMello characterises the practice as
translation; that is putting into human words the minds of non-human animals
(DeMello, 2013: 5). Implicit in Taylor's questions, then, is the ongoing issue of
who is authorised to translate and what type of human language is needed for such
descriptions. Certainly, such an endeavour carries a huge ethical weight and there
is a certain irony to the situation where, due to the charge of anthropomorphism,
the human language used to describe animal language is so heavily contested.

The human-animal politics of language

The linguist Ferdinand de Saussure argued that the 'speech circuit' must be composed of two individuals that share the same signs linked to the same concepts (de
Saussure, 1915). De Saussure posited that language should be considered to be
composed of 'langue' and 'parole', wherein 'langue' was the system of language
rules and parole was the manifestation of those rules within speech or writing.
For some, nonhuman animal language is rendered problematic by de Saussure's
model of the 'speech circuit' for a number of reasons; first, the biological difference between humans and other animals creates a point of contention because,
quite simply, nonhuman animals do not possess the same capacity to 'speak' or

'write'; in the second place, the speech circuit implicitly assumes that the communication of shared concepts between 'speakers' is enabled by the minds of the speaking agents. In this sense the attribution of selfhood and mind is intimately connected to the attribution of language which presents a double bind in terms of human-nonhuman animal relations and the anthropocentric preference for asymmetries of power. In short, issues of 'animal minds' and the moral status of animals are intrinsically connected to the attribution of language. Language and mind are thus resolutely bound together with the question of the rights of a speaking agent.

Despite scientific arguments that favour the case for human exceptionalism based on language, there is little evidence in popular culture that such a discourse has quite the same traction in the public consciousness. Indeed, as one linguist writes:

> It seems fair to say that the current understanding in the popular press is that the conception of language as an ability limited to humans is not only outmoded but even a kind of prejudice that science has shown to be wrong.
>
> (Anderson, 2006: 3)

There is no doubt that there have been scientific studies on animal language and communication which have attracted major press attention since the mid-twentieth century and that many of these have become the influence or at least the loose basis for fictional literary, film or television narratives. John Lilly's experiments with dolphin-human communication began in the late 1950s and his ideas were expanded in the popular book *Man and Dolphin* in 1961. Lilly's work (which is discussed later in this chapter) directly influenced the novel and later film adaptation *The Day of the Dolphin*, the television series *Flipper* and the film *Namu: The Killer Whale*. In the late 1970s, Irene Pepperberg began the 'avian language experiment' with an African grey parrot called Alex. Their relationship became the basis for a script for a family comedy film titled *Paulie* (1998), a story about a highly intelligent talking blue-crowned conure. A less financially successful Australian film about a talking macaw, titled *The Real Macaw* (1998), was also released the same year.

The discovery by Roger and Katy Payne and Scott McVay that humpback whales created songs led to the production of the 1970 LP vinyl record *Songs of the Humpback Whale*. The album sold more than 100,000 copies and parts of the recording were included in the Golden Record carried on the *Voyager* spacecraft launched in 1977. As Paul Spong writes, the purpose of the Golden Record was to take evidence 'of the diversity and complexity of life on Earth to the far reaches of our solar system' in the hope that other beings would know that our planet 'had nurtured intelligence; the languages of human culture were there and so were the voices of whales' (Spong, 2011: 131–132). D Graham Burnett writes about the humpback whale recordings that 'It would be difficult to overstate the significance of this work to end whaling' (Burnett, 2012: 629). They became, he writes 'nothing less than the soundtrack of the "Save the Whales" campaign' (ibid.). Yet, scientific discovery of nonhuman animal communication is not in itself enough to ensure public empathy for a species. In 1973, the zoologist Karl von Frisch was given the Nobel Prize for his discovery that bees were able to communicate the location of food sources to each other by engaging in behaviour known as the waggle dance. Unlike the fictionalised reimaginings of dolphin and ape studies, this discovery did little to change public attitudes in North America towards bees following continued reports after 1957 about 26 African queen bees who had escaped from Brazil. The bees, having interbred with domestic Brazilian bees, were referred to as 'aggressive' and reports that swarms had 'killed animals and humans' fed public anxieties, which were reflected in a cycle of horror films (Rensberger, 1972; *New York Times*, 1978). While scientific accounts varied as to when the hybrid bees would migrate to North America, popular culture imagined the outcome when the swarms did eventually arrive. Von Frisch's discovery of bee communication permeated into some of these popular imaginings with deadly consequences, for instance through human communication with, and control of, bees and the collective intelligence of swarms. *The Deadly Bees* (1967), *Killer Bees* (1974), *The Savage Bees* (1976), *The Bees* (1978), *Terror Out of the Sky* (1978) and *The Swarm* (1978), an adaptation of Arthur Herzog's 1974 book of the same title, coupled with the press reports shaped the public thinking about bees to such an extent that in 1978 the US Agriculture Department produced a 14-minute film to offer reassurance that the bees were manageable and referred to them as a 'much-maligned insect' (*New York Times*, 24 July 1978). The cycle of bee horror movies had exhausted itself by the end of the 1970s, and although there was a brief resurgence of interest in bee attack movies with *The Deadly Invasion* (1995) and *Killer Bees!* (2002), widespread reporting of colony collapse disorder in the mid-2000s, films such as *Bee Movie* (2007), *Maya the Bee* (2015) and the documentary films *Vanishing of the Bees* (2009), *More than Honey* (2012) and *Queen of the Sun* (2010) reflected the vast shift in public perceptions of bees from those of the 1970s. Significant in terms of challenging previous discourses on bees, signifiers of vulnerability replaced those of aggression as affective appeals that mobilised public concern.

By the 1980s the specialist field of bioacoustics contributed much to the scientific debates about nonhuman animal language and communication. In 1984, Katherine Payne argued that elephants could communicate with each other over large distances using infrasound that operates below the level of human hearing and the study of whale and dolphin communication at research centres such as the Bioacoustics Research Program at Cornell Laboratory of Ornithology demonstrated how marine mammals interacted socially. Whilst earlier research such as the Washoe project had been met with much scepticism within the scientific community, by the 1980s the study of nonhuman animal language and communication was established as a credible, if still controversial, area of scientific research. Hollywood films, both animated and live action, were by this time, overrun with nonhuman animals who were bipedal, clothed, talking, singing, dancing zoomorphs of one type of another, and narratives of interspecies communication were a major box office draw.

Talking primates

Early experiments to teach a chimpanzee called Washoe American Sign Language (ASL) were conducted in the late 1960s after previous attempts to teach primates language and vocalisation had failed. The Washoe project, while controversial, foregrounded the possibility that humans were not unique in their use of language. Before Washoe, in 1909 there was Peter, a young chimpanzee who, psychologist Lightner Witmer argued, had sufficient intelligence to make worthwhile experiments to teach him to speak. In 1933 Winthrop and Luella Kellogg began experiments to raise a young chimpanzee called Gua as a child and teach her to speak, and two decades later Keith Hays and Katherine Hays began to report on their experiments to teach a chimpanzee named Viki to speak. Despite the lack of success with the earlier experiments, Hollywood had, by the 1940s, already seized on the idea of chimpanzees communicating with humans and used it as the basis for comedy moments in the Tarzan series of films. For example, at the end of *Tarzan Triumphs* (1943) Cheeta takes a shortwave radio microphone and his vocalisations are mistaken for Hitler by Wehrmacht officers. By the 1950s, chimpanzees' intelligence and their capacity for language acquisition was the central premise of films such as *Bedtime for Bonzo* (1951) and *Bonzo Goes to College* (1952). Other films such as *Monkey Business* (1952) and the pairing of Jerry Lewis with a female chimpanzee named Pierre in *My Friend Irma Goes West* (1950) capitalised on the public fascination with chimpanzee-human similitude. Press reports of scientific studies that talked about equivalence between children and chimpanzee intelligence and chimpanzees' capacities for communication were regularly leveraged by Hollywood studios and used in the marketing and promotion of the animal star personas (see Molloy, 2011). Further experiments in language acquisition in chimpanzees began in 1973 with another individual named Nim Chimpsky. Project Nim attempted to recreate the Washoe project and was funded until 1977. In 1980 Sue Savage-Rumbaugh began work with bonobos at the Georgia State University Language Research Centre. The main subject of Savage-Rumbaugh's work was a bonobo named Kanzi who was taught to use lexigrams, a system of communication that uses symbols to represent words. In the 1970s Penny Patterson began teaching a gorilla named Koko sign language. In 1978 Koko was featured on the cover of *National Geographic* and was the subject of the documentary *Koko: A Talking Gorilla* by Barbet Schroeder. Koko's story became the inspiration for Michael Crichton's best-selling novel *Congo* in 1980.

Talking chimpanzees as objects of humour continued to populate popular culture. US television series *Lancelot Link, Secret Chimp* (1970–1971), a *James Bond* parody, and a series of UK television advertisements for PG Tips tea which ran from 1956 until 2002, used chimpanzees dressed in clothes and overdubbed their mouth movements with human voices. Humanised 'ape sidekicks' and the performance of interspecies communication, usually exploited for comedic 'value', were normalised through frequent appearances in popular culture, for example *The Today Show, BJ and the Bear* and the films *Every Which Way But Loose* (1978) and *Any Which Way You Can* (1980). Chimpanzees in human clothes and settings were not a relic of twentieth-century exploitation, and individuals contin-ued to appear in twenty-first-century films such as *The Wolf of Wall Street* (2013) and *The Hangover Part II* (2011). The popular culture imaginings of nonhuman animal language have done much to shape public perceptions of the communica-tive capacities of particular species. In the case of apes, there is no doubt that the ubiquity of 'talking chimps' has had disastrous consequences and played a role in fuelling fundamental misunderstandings about human-chimpanzee similitude and the suitability of chimpanzees as companions, masked the cruel exploitation of chimpanzees in the entertainment industries and contributed to a distorted public understanding of threats to species (Molloy, 2011, 2012; Aldrich, 2018).

In 2012, the Walt Disney Company published a policy on its use of live animals in entertainment which expressly prohibited 'the use of apes (chimpanzees, goril-las, orangutans, bonobos, gibbons and siamangs) or other large primates (baboons and macaques) outside of their zoo/sanctuary habitat or natural environment' (Dis-ney, 2012). It is important to note that Disney begin the policy wording by point-ing out that the brand 'has a rich heritage of including animals in its entertainment experiences' (ibid.). While such moves can be seen to reflect corporate concern for the wellbeing of the species mentioned, it is also the case that the brand's asso-ciations are such that Disney must protect its reputational capital (Molloy, 2011). Thus, it is not unrealistic to suggest that corporate responsibility for the welfare of (certain) animals is underwritten by pressures to protect the public perception and currency of a brand. The same year that Disney produced its policy on live animals, the company's nature documentary unit, Disneynature, released the film *Chim-panzee* (2012). The film followed a free roaming young chimpanzee named Oscar and although described in critical reviews as 'relentlessly' anthropomorphic and 'trivialising' the life of the young chimpanzee (Lumenick, 2012), it also received praise for being 'an eye-opening primer in cross-species similarity' (Williams, 2012). The affective appeals of the film were widely remarked on as critics noted that the norms of chimpanzee life were sanitised for younger viewers, although this was not universally acknowledged as a positive and some critics commented that the narration amplified 'cuteness' at the expense of the chimpanzees' dignity (see Catsoulis, 2012). A donation from each ticket sold in the opening weekend of the film's release was made to the Jane Goodall Institute, which had been active in raising awareness about the exploitation, suffering and consequences of featuring chimpanzees in entertainment and advertising (The Jane Goodall Institute, n.d.). Reassuring audiences of the credibility of the film, *Chimpanzee* followed the typi-cal Disneynature marketing and release strategy and aimed to reinforce 'Disney's overall strategy to associate its brand with an environmentally sustainable busi-ness model, conservation and environmental protection' (Molloy, 2013: 174). The interior monologue of chimpanzees was imagined and narrated for the feature film while *Oscar's Chimp Diaries*, a series of short videos designed for social media sharing, featured Oscar narrating his own experiences in the forest, voiced by a young male human. This presents a complex site of anthropomorphism, one where the emotional labour of animals is realised through the narrative's affective appeals while the commodification of anthropomorphism may arguably distance humans from the realities of chimpanzee life in a form that is commercially sale-able to family audiences. At the same time, the anthropomorphism of *Chimpanzee* emphasises the emotional lives of chimpanzees. This also happens in a forest and not a human setting that would have previously been favoured for a 'chimpanzee story'. Moving the stories of chimpanzees from houses to forests exists in a com-mercial context that draws attention to the commercial pressure to maintain brand associations with welfare concerns and protect reputational capital. This situation also reveals an inherent vulnerability for industries that rely on the exploitation of animals and their labour and encourage the investment of human consumers in anthropomorphism as a means to sell the product. Sites of commodified anthropo-morphism and the emotional labour of animals are vulnerable to public scrutiny of a company's moral regard for the animals due in part to the anthropomorphism that the consumers have been encouraged to emotionally invest in. However, in an era of prosumer capitalism, it is not only large commercial companies which exploit the affective labour of talking animals.

Talking dogs and talking with dogs

Scientific studies of animal communication have continued to garner plenty of press attention but there are other reasons that may account for the general public acceptance of the idea that nonhuman animals possess language. First, the distinc-tions between language and communication that might endlessly absorb philoso-phers and scientists are not keenly debated in the press discourse where stories about talking nonhuman animals are newsworthy, even though media outlets will adopt different reporting strategies, as evidenced by the articles about NOC. Sec-ond, there is no shortage of talking animals in popular culture and despite, or perhaps due to their ubiquity there still exists a general fascination with the notion of interspecies communication, albeit not exclusively between humans and other species. Third, talking to a nonhuman human animal is normalised within human-companion nonhuman animal relationships. Studies have shown that humans consider nonhuman animal companions to be members of the family (Beck and Katcher, 1996; Charles and Davies, 2008; Charles, 2016) and a majority (around 90 per cent) talk to them every day (Pierce, 2016: 115). Indeed, a touch-talk rela-tionship between human and nonhuman animals is considered unavoidable to the extent that it is 'almost impossible for pet owners to touch their dogs or cats without talking to them at the same time' (Friedman, 2013: 48). Those who live with birds reportedly show an even greater propensity for talking with their avian companion than those who live with dogs or cats. Moreover, talk is considered to be an 'essential component' of human-companion nonhuman animal relation-ships because it 'creates companionship', which has been claimed to have a posi-tive influence on human health (Beck and Katcher, 1996: 84). Clinton Sanders and Arnold Arluke argue that those who live with companion animals view them as 'virtual persons', minded conscious, purposive coactors who 'engage in com-municative activities similar to those of the other family members'. They claim that 'it is through ongoing interactional experience with the dog that the owner learns to "read" gaze, vocalizations, bodily expression, and other communicative acts' (Arluke and Sanders, 1996). Anthropomorphism, according to Arluke and Sanders, has a role to play in such relationships where, they argue, it can function as a useful heuristic device' (ibid.). In other words, communication that supports social interaction and bonding between human and companion nonhuman animals does not require both individuals to have the capacity for language.

There are different types of talking in the companion nonhuman animal-human relationship. Arluke and Sanders note the different modes of 'speaking for' dogs that include talking on their behalf at veterinary appointments, constructing dialogue-like exchanges and presenting the 'virtual voice' of a dog to express the speaker's own wants or concerns (1996: 67–70). The act of speaking for a dog, they argue, demonstrates intimacy and actively constructs the identity of the nonhuman animal (ibid.). These forms of 'speaking for' are thus attempts by human coactors to give a voice to the subjective experience of their canine companions. We can also add to the acts of nonhuman animal speaking, the 'cutified' performance of communicative relationships between humans and canines which pervade social media.

Viral 'talking dog' videos reify communicative relationships such that the clips are a recognised genre in their own right, often transformed by commercial enterprises into compilations that might run for eight to ten minutes. Viral 'funny' videos which depict dogs being encouraged by their human companions to produce sounds that loosely approximate the words 'I love you' are, Karla Armbruster observes, a 'grotesque performance' that does not respect canine difference and tells us what we want to hear: dogs confirming that we are 'inherently lovable' (Armbruster, 2013: 17). Usually one continuous shot and set in the domestic space, in these clips a human stands over a dog, who looks up and into the camera. The human repeats the phrase 'I love you' in a high-pitched tone while the dog makes the various yips, barks, growls and grunts that eventually result in praise, a food treat or the cessation of the human demand to 'talk'. Framed as comical, as Armbruster suggests, these videos may provoke intense discomfort in anyone concerned about canine difference. In compilation videos this discomfort is likely magnified by the continuous stream of clips edited together which make apparent an aesthetic convention of domination in which humans stand over their canine companions repeatedly screeching 'I love you' until the right combination of dog vocalisations is 'returned'.

The public reception of these videos, reflected in their viral status, reactions and comments, tells an interesting story about the various ways in which social media users engage with this type of content. In the case of one 'funny talking dog' compilation video on YouTube, which was viewed over 868,000 times, perhaps unsurprisingly the majority of comments described the video primarily as funny (22.3%) cute (18.5%) or generally expressed 'love' for the video and dogs (11.8%).[2] Other comments, while in the minority, suggested that the videos were disrespectful to the dogs (4%); that they demonstrated the intelligence of dogs (6%); that they showed a breed-specific trait (6%); or that the commenter knew or lived with a dog who did something similar (4%). A small number of comments (only 2%) expressed a desire to truly understand what dogs are trying to communicate, while 8% of comments 'translated' what the dogs were saying, re-scripting the dogs' utterances in the form of a parody.

Titles are crucial framing devices that place the videos within a predetermined discourse of humour and direct their reading to such an extent that those who post comments which challenge the comic value are inevitably criticised for being killjoys. On the face of it, these videos celebrate the performance of dogs as human imitators, as caricatures who can reflect back the love necessary to satisfy our need for validation. They demonstrate a human proclivity to ignore other subjectivities in favour of our own desires and we can question whether the spectacle of 'talking dogs' in popular culture is a harmless amusement or a reminder of the many dimensions of our anthropocentric narcissism. There is certainly a case to be made that these performances of interspecies communication are a parody of empathetic conversations with other animals. These dogs are granted a type of managed agency within a form of anthropocentric propaganda that fosters the persuasive illusion that other animals have value because they amuse us with their impoverished attempts to be human-like. This may be particularly true of compilation 'talking dog' videos that have been curated by commercial companies, a form of reification by remediation. The social relations between human and companion nonhuman animal are instrumentalised and 'thingified' in these ten-or-so-minute compilations and have value precisely because they function as a series of reductive 'cute' comedy moments strung together to occupy the focus of a viewer long enough that they will also watch the four or more five-second advertising spots dispersed along the timeline. If, as Sianne Ngai argues, the cute commodity makes powerful affective demands (2012: 64), the cutification of talking dogs and the sharing of these communicative performances reveals how asymmetries of power are reproduced through reification of the emotional labour of nonhuman animals.

Predecessors of these compilation videos are a variant of reality-based television programming that use clips from home video sent in by viewers. Programmes such as *You've Been Framed* (UK), *America's Funniest Home Videos* and *Australia's Funniest Home Videos* were the forerunners of the compilation videos although the social logics and cultural practices associated with what Henry Jenkins terms 'spreadable media', differ from the traditional models attributed to broadcast media (Jenkins et al., 2013). Spreadable media is media that is circulated across platforms through the actions of networks of consumer/participants. The YouTube platform has low barriers to entry and supports a wide variety of users who have different motivations for uploading content. The aggregation of attention, building of notoriety, communal sharing, information gathering, education and archiving are all noted by Jenkins, Green and Ford as driving the uploading of content to the platform (2013: 93). Content, at an individual level, is appraised and the assessment of the value of texts 'as a resource for social exchange' as well as 'sentimental value and personal interest' inform the decisions to circulate content more broadly. The aggregation of these individual decisions to circulate 'may help determine the economic value of a particular video, assisting media companies in mapping large-scale patterns of taste and interest that may cut across multiple social networks' (2013: 95). This understanding of aggregated individual content circulation can then be exploited to generate revenue. Companies sift through hundreds, even thousands, of hours of user generated clips to identify shareable content, do deals with the content owners and then license the clips to third parties or compile them into longer genre-based videos with embedded advertising. The business of viral video has resulted in the transformation of some content creators and nonhuman animal co-creators into celebrities and the formation of a distinct sub-genre of 'talking dog' video clips that have an identifiable set of codes and conventions. This is not to say that talking dog videos are a product of social media.[3] However, the ways in which the content is produced and distributed, how users interact with that content and the means by which 'cute' appeal drives the fetishisation of certain aspects of interspecies communicative relationships have all been transformed.

Cute

'Cute cultures' are part of an expanding and pervasive global phenomenon of cuteness which has previously been trivialised. This trivialisation belies the power dynamics of cuteness, the construction of which Matthew Cole and Kate Stewart argue relies on a synthesis of infantilism and aesthetics that does not trouble instrumental human-nonhuman animal relations (Cole and Stewart, 2014: 101). As Joshua Paul Dale et al. argue, cuteness is 'more than a facile commodity aesthetic'; it has serious implications in relation to the power differentials 'between a subject affected by cuteness and an ostensibly powerless cute object' as well as providing, they argue, 'an important coping strategy for subjects caught up in the precariousness inherent to neoliberal capitalism, and is thus central to the establishment of contemporary (inter)subjectivities' (Dale et al., 2017: 2). Offering an explanation for the proliferation of 'cute' nonhuman animal videos which occurred in tandem with the rise of austerity and the corporatisation of everyday life, Allison Page contends that cuteness is reassuring and offers affective experiences and connections that temporarily salve the deleterious effects of neoliberalism (Page, 2017). 'Cute animal videos', Page argues, 'are imagined as powerful vehicles through which humanity is revived, consoled, and/or healed from the suffering of the day' (Page, 2017: 88). While 'cuteness' may offer a momentary respite from the anxieties of contemporary life under neoliberalism, the cultural and affective currency of 'cute' talking dog videos is simultaneously exploited for its economic worth by the same neoliberal logics. Indeed, it is of note that recent studies have focused on the general increase in the use of cute animals for marketing purposes and the extent to which cuteness invites positive affective feelings of care whilst also being linked to indulgent consumer behaviour (Nenkov and Scott, 2014).

Cuteness is often considered to be akin to anthropomorphism where cute is taken to be a response that is triggered by the *Kindchenschema* ('child schema'). This theory has spawned one argument that 'pets are simply social parasites who have perfected the art of releasing and exploiting our innate parental instincts'

(Serpell, 2003: 87). However, James Serpell argues that, in the case of puppies and dogs, this presents a disparaging theory of the companion relationship and he argues that 'people may indeed find puppies or Pug Dogs cute, but they certainly are never in any doubt concerning their true provenance' (Serpell, 2003: 87). In this sense, Serpell proposes that humans are not being duped by the 'cute response' into caring for other animals. Not only are they aware of what the relationship entails but humans reap social, mental and physical benefits from living with a companion nonhuman animal. The *Kindchenschema* that Konrad Lorenz (1943) argued activates an innate human motivation for care has been challenged in recent scholarship where it is argued that cuteness is better understood as an appeal 'that seeks to trigger an affective response' (Dale et. al, 2017: 5). The cute aesthetic of talking dog videos is constructed primarily by reducing individual dogs to childlike partial subjectivities in a performance where the human adopts the role of the accomplished adult with full capacity for speech, teaching and encouraging the infantalised canine in their imperfect attempts at emulation. In these videos, humans use 'baby talk' characterised by a high-pitched tone, repetition, slow and clear enunciation and restricted vocabulary. By being filmed in the home or sometimes in cars, talking dog videos are differentiated from the manufactured performances of animal 'actors', the setting providing signifiers of authenticity that exaggerate the combined whimsy, familiarity and normality of the moment. Conventions such as standing above and looking down with the camera at an individual dog not only magnifies the asymmetry of power but actively 'makes little', reinforcing notions of the pseudo-child and undercutting potential de-cutifying aspects such as life stage (adult dogs are 'less cute' than puppies), breed characteristics, or other individual physiognomic features. The act of 'making little' diminishes both the physical presence of the dog as well as their animal subjecthood. Cuteness is thus relational, relying on the imbalance between the diminished weaker object and the larger stronger subject, a power asymmetry that must be observed in order for the cute object to activate an affective response in the viewer (Ngai, 2012).

Cuteness invites care through the affective draw of inequalities of power. In relation to talking dog videos there are, however, interesting corollaries of cuteness that bear consideration and are perhaps illustrated through the example of Mishka, a Siberian husky and internet celebrity. With more than 867,000 subscribers to her YouTube channel, over 500 million video views in nine years, over one million Facebook likes and more than 990,000 followers, Mishka was well-known on social media as 'Mishka, the Talking Husky'. The first video of Mishka, posted in 2008, was a 48-second clip entitled 'Husky Dog Talking – "I love you"' and was viewed over 102 million times.[4] Between 2008 and 2017, 719 videos of Mishka were uploaded, some of which also featured other dogs or Mishka engaged in another activity, but mainly the clips focused on her vocalisations. Where Mishka was vocalising freely and not depicted as mimicking human cues, that is to say, when she was not typically being encouraged to say 'I love you', subtitles were used to translate her various howls and growls. Mishka appeared in television commercials, films, in news items and on television talk shows and was central to the emergence of the talking dog video phenomenon on social media. On the announcement of her death in April 2017, over 17,000 comments were left on Mishka's Facebook page. The comments reflected one or more of the following categories of response: expression of condolences for Mishka's human companions; an expression of a close personal association/bond with Mishka as a 'virtual pet'; statement that Mishka's videos motivated the adoption from rescue or acquisition of a husky; shared experience of grief over the loss of a companion nonhuman animal (another dog or specifically a husky); recollections of affective responses to the videos, typically 'joy', 'delight', 'happiness', 'smiles'; noted emulation of Mishka's 'I love you' by and between humans (parent and child, partnered couples); the inclusion of Mishka's vocalisations ('awuvu') as part of a human language of intimacy; and belief in an afterlife (heaven, Rainbow Bridge, etc.) for dogs. The consumption of cuteness as a salve for the precarity of contemporary life is reflected in the recollections of affective responses to Mishka's videos, for example, 'she put countless smiles on my face when I was having a bad day or just needed a pick-me-up', 'watching all your videos brought a smile to all our faces and warmed our hearts' and 'I felt like a part of your extended family. Whenever I was sad or lonely, I'd always watch your videos'. It is indeed the case that increasingly nonhuman animals are valued not for their 'practical' use but for their emotional labour. In the case of talking dog clips, this emotional labour is constructed and normalised by the cutification of canines in this genre and enabled through the sharing and interactions particular to social media platforms and practices.

The emotional labour of nonhuman animals generally can mobilise empathetic concerns and action. Many who commented on the announcement of Mishka's death claimed that watching her 'talking dog' videos had led them to adopt a husky from a rescue organisation: 'your videos of Mishka originally inspired me to go out and get my own husky'; 'she was so adorable inspired me to get a husky/malamute mix from the shelter'; and 'Mishka was my entry into the fabulous world of huskies'. This suggests that the emotional pull of cute nonhuman animal media does travel beyond the affective flows of the internet and acts as a significant driver, particularly in relation to 'pet culture'. There is thus an inescapable connection between affect and empathy that is expressed in the case of talking dogs as a relation between cuteness and compassion. Certainly, the cutification of talking dogs relies on a performance of passivity, a diminishing of animality which is replaced by infantalised humanisation that in turn creates a desire for intimacy. However, at the same time, the capacity to 'talk' is considered to be a husky-specific trait and, far from diminishing a dog's subjecthood, it is constructed by those who have experience of living with huskies as individuating, a key aspect of their personhood and mindedness, and evidence of the breeds' collective ancient ancestry. As the author of one guide to the breed remarks:

> Siberian Huskies, true to their heritage, bark less than many other breeds. When they do vocalize they tend to woo, howl, sing, chirp, and chatter softly. [. . .] The plaintive 'wooing' sound is a Siberian trademark. [. . .] Howling is a primitive

trait in dogs, which means that breeds of ancient lineage are much more likely
to howl than more modern breeds.

(Morgan, 2001: 78–79)

Discussions in online Siberian husky companion communities customarily
make references to the vocalisations of individuals and accounts of the experiences of learning the subtleties of interspecies communication between human
and husky are common. Human companions describe the meanings of vocalisations which are accepted to be specific to the individual dog. For example, one
forum member writes 'I THINK I am starting to understand what he is saying . . .
but I would imagine all our huskies are probably different with their "meanings"'
(DogForums, 2012); another advises a first time 'owner', 'after a while, you'll be
able to know what each noise means' (DogForums, 2012); and there are whole
threads devoted to huskies who 'talk back' or 'complain' when asked to do something they object to. In this way, what is constructed as anthropomorphic cutification reliant on the diminishing of animal subjecthood in one context carries quite
different meanings in everyday husky-human relations. In this case, the experience of interspecies communication between husky and human is an enhanced
version of the talk relationship that Arluke and Sanders note where individual
dogs are constructed as minded conscious, purposive coactors whose subjecthood
is understood, in part, through their engagement in communicative activities.

Comedy value and talking nonhuman animals

Talking nonhuman animals have affective value and intertextual references are
often utilised to trivialise interspecies communication, a means by which popular culture's imaginings of nonhuman language are invoked to buttress anthropocentric boundaries. For instance, the 1995 blockbuster *Congo*, an adaptation
of Michael Crichton's 1980 book of the same name, liberally references other
talking nonhuman animal characters in popular culture to reflexively signal that it
is a self-aware comedy action movie about a talking nonhuman ape. In the film,
Amy, a gorilla, can communicate by using assistive technology in the form of a
backpack that contains a voice simulator and a glove which interprets her use of
American Sign Language. 'Animals can talk', asserts Peter the primatologist, in
front of a half empty lecture hall. Amy, clutching her favourite teddy bear – a
signifier that we are expected to read Amy as infantalised and therefore cute –
responds to Peter saying 'Amy seven', 'Amy good gorilla' and 'Amy pretty'.
As Amy speaks, a man in the audience leans towards the woman in the next seat
and says 'this is a talking gorilla Moira. This gorilla is talking. But this is really
happening. This isn't Mr Ed'. Indeed, Amy is not a Mister Ed–type of character. Instead she is a cyborg, an animal-machine, enhanced by technology for the
purposes of interspecies communication. Without the assistive technologies and
voice simulator, Amy is silenced.

Technologically enhanced animals with the capacity to communicate with
humans are a common feature in popular culture. *The Cat from Outer Space,*

a live action film, features a cat with a collar that allows him to communicate; Dug the dog, a character from the feature-length animated film *Up!* and the Pixar short *Dug's Special Mission*, also wears a collar that translates his thoughts into human speech; Jones the dolphin in the 1995 film *Johnny Mnemonic* is connected to an array of communications and computer technologies and is an expert code breaker; and Rocket Raccoon, the Marvel superhero who features in the *Guardians of the Galaxy* films, has been genetically manipulated to grant him humanlike intelligence and speech. Like Amy the gorilla, each of these nonhuman animal characters is enhanced by technology to enable interspecies communication on wholly human terms. The film *Congo* questions the ethics of making other animals talk when it becomes apparent that Amy's capacity for human speech is an obstacle she must overcome to be accepted into a wild gorilla group. In being 'more than' an ape, Amy's gorilla-ness is diminished, and it is only when she uses her own species-specific gorilla communicative strategies does she save Peter and become accepted into the gorilla group.

Congo plays with the clichés of the jungle adventure story genre and recalls the various adaptations of *King Solomon's Mines*, the original of which (by H. Rider Haggard) Crichton had used as the influence for his 1980 book. In the film, Amy the talking gorilla is a useful plot device to get a group of humans into a jungle setting led by Peter's decision to 'release her back to the wild'. Amy is an intentionally comic character played by a human in an ape suit. She sips a martini, smokes a cigarette and is jealous of the main female character, who she describes as 'ugly'. Her dialogue is infantalised and the audience is left with no doubt that the talking gorilla onscreen is not to be taken seriously. Yet, the original novel, *Congo* (1980), was inspired by the story of Koko the gorilla, and by Crichton's interest in the implications of sign language research for animal rights. Commenting on the critical reception of his book in 1980, Crichton wrote, 'When the book was published, most reviewers found the character of Amy, the sign-language-using gorilla, too incredible to believe. This despite the fact that I had modelled Amy on a real signing gorilla, Koko, then at Stanford University' (Crichton, 2015). Although she had been on the covers of *National Geographic* magazine and *New York Times Magazine* and was, in Crichton's view, 'pretty famous', he noted, 'apparently, book reviewers had never heard of Koko' (ibid.). The film, produced 15 years after the critical reviews of the original book, managed the incredulity of critics by adapting the story of a talking gorilla as a comedy and downplaying questions about the ethical implications of interspecies communication.

There are many such examples of comedic talking nonhuman animals that populate live action and animated worlds. In the case of *Mister Ed*, a television series adapted from the short stories of Walter Brooks and originally broadcast from 1961 to 1966, the comic premise of the show is that the horse talks to only one person, Wilbur Post. Throughout the series, Wilbur's mental state is brought into question whenever he mentions, usually accidentally, that Mister Ed speaks to him. In the original short stories, the human character, Wilbur Pope, and Mister Ed are usually drunk, but the television show had to take a different, more family friendly and therefore sober approach to the relationship. In the first episode,

Wilbur is hit on the head by a rake that is lying conveniently on the ground in the yard of the new house he has just purchased with his wife, Connie. From that moment on, he is able to converse with the horse who has been left behind by the previous owners of the house. The head injury in the television series and the drunkenness in the original stories give licence to the notion that the character of Wilbur is an unreliable narrator, a trope shared by the character Elwood P. Dowd in the film *Harvey* (1950). Dowd is a wealthy drunk who talks with a 6'3.5" rabbit called Harvey who only Dowd can see and hear. Mister Ed and Harvey, both talking animal characters, have in common a mischievous personality and a refusal to speak to anyone other than the main male character. In *Harvey*, Dowd reveals that the eponymous nonhuman animal character is a pooka, a fairy spirit from Celtic mythology which usually takes the form of a horse, ass, goat, dog, eagle, or (in this case) a rabbit. A pooka is a trickster that can speak and prefers to pick on drunkards who won't be believed when they recount their tale the following day. Although not made explicit in the *Mister Ed* series, the horse character has all the traits of a pooka as does Francis the Talking Mule – another talking equine who appeared in seven feature films between 1950 and 1956, again only conceding to talk with one male character, a young soldier. In the *Francis* films, the mule continually engineers situations where the soldier is forced to admit to those around him that Francis can talk. Consequently, the soldier's mental health is repeatedly called into question. There is thus a recurring theme in stories of talking nonhuman animals: while the male characters in the *Mister Ed* television series and *Francis* films were not depicted as drunkards, their mental state comes under scrutiny due to the actions of the puckish equines only they are able to converse with. These talking nonhuman animals, having an excellent command of language, are depicted as having greater intelligence than their human counterparts. They lack the characteristics of 'cuteness' in that they are self-sufficient, their vocabulary is extensive, they have deep masculine voices and they are able to manipulate situations using their advanced capacity for human language. In this regard they occupy a different category of anthropomorphic 'talkers' to Amy and indeed Koko the gorilla, or some of the talking dogs who populate social media. Accomplished talkers do not elicit the emotional appeals that infantalised talkers can evoke. Thus, at certain sites the infantalisation of speech can operate as a compelling signifier of cuteness that is framed by unequal power relations and the diminishing of animal subjecthood. Nonetheless, the whimsy of talking nonhuman animals allows an adult audience to indulge in the pleasures of such narratives, whether cute or not.

Talking animals in science and fiction

Where things become a little more complex perhaps are the sites where science and fiction are interwoven. In the case of talking nonhuman animals, this is apparent in the cases of marine mammals, particularly whales, who have been the focus of much scientific interest with the results of language experiments widely disseminated by the press and used as the inspiration for a number of popular literary, film and television narratives. NOC used human language for four years, from 1984 until 1988, after which he communicated only in beluga. It was noted that NOC's cessation of communication with humans coincided with his reaching sexual maturity. When asked why it took until 2012 for the scientists to write an article about NOC's vocalisations which occurred in the 1980s, the lead author claimed he had thought that 'a talking whale was a "side issue"' (Joyce, 2012). As a twenty-first-century online news story, however, a talking whale was a major attraction, and some journalists skimmed over the issue of the 30-year gap to present the events as if the discovery had just been made. 'Scientists have found a white whale capable of imitating human speech' (Choi, 2012), one publication declared, as if NOC was still alive. Another similarly retold the story as if the event had only recently occurred: 'A captive white whale that made unusual mumbling sounds when he was in the presence of people may have been trying to mimic his human companions, scientists have found' (Connor, 2012). By way of an explanation for the lag in reporting NOC's vocalisations, the article noted that the scientists had only now 'analysed the archived sound recordings made when NOC was alive' (ibid.). Other online reports were sceptical about the timing of the research paper which was published four months after a public outcry over an application to import 18 wild-caught beluga whales from Russia. Some argued that the article was part of a publicity stunt to undermine opposition to the importation plans and garner public support for captive research programmes (Batt, 2012). It is also salient to point out that the problems associated with John C. Lilly's highly controversial work on dolphin-human communication continued to impact on how scientists discussed interspecies communication in the 1980s. Accounts of a talking beluga whale would not have been favourably received by many in the scientific establishment keen to put distance between Lilly's work and, what was regarded as, serious scientific enquiry. What was certainly absent from the majority of the 2012 media coverage of NOC's vocalisations was that he had been part of the Navy Marine Mammal Program (NMMP) and captured for the purposes of a naval initiative known as 'Cold Ops'.

The NMMP used dolphins and sea lions since the early 1960s but as cold war tensions re-focused activities around the Arctic, the program had to find other species that would be able to survive the extreme environment. NOC, captured on 4 August 1977 in Hudson Bay, Canada, was one of six beluga whales used by the NMMP between 1977 and 1980. Of the six captured belugas, Muktuk was transferred to SeaWorld San Diego in February 2001, where she died in January 2007; Ruby, another female caught in 1980, was also moved to SeaWorld where she died in July 2014 of fungal encephalitis; Churchill, a male, stayed in the program until his death from pneumonia in 1985; NOC died in April 1999 of Aspergillus encephalitis; a third female, known as LLY, only survived for two years in the program, dying from drowning in November 1982; and a female named CHR lived only two years longer than LLY, dying on 14 July 1984, almost four years from her capture date on 16 July 1980.[5] All six were captured by Inuit hunters hired by the US Navy. Belugas are known for being friendly and curious which, according to one military journalist, made NOC's capture 'straightforward'. One hunter would, he wrote, 'jump on its back and lasso it. Another Inuit would slip a stretcher underneath the whale and move it to shore'. The process was described by a marine mammal trainer who assisted with the capture as resembling 'a rodeo', following which 'each whale was placed in a watertight box, lying on a stretcher in a few centimetres of water. It was loaded on a military transport jet to be taken to the United States' (Pugliese, 2003).

The lead author of the *Current Biology* article co-founded the NMMP in 1961 and moved, in 2007, to the role of director of the National Marine Mammal Foundation, an organisation which states on its website that it 'proudly helps to care for Navy marine mammals'.[6] Although the relationship between the NMMF and the NMMP is not made explicit, the NMMF's site notes that 'over the past half-century, the Navy Marine Mammal Program – including many NMMF scientists – has been the largest contributor to marine mammal science' (NMMF). Due to the lack of explicit links between the NMMF and NMMP, when the original article on NOC was published with an NMMF affiliation, the realities of NOC's life as a 'marine-mammal system' were obscured.

Historically, dolphins were favoured by the navy programme, initially for hydrodynamic studies but by 1962 the focus had turned to their echolocation capabilities. By the mid-1960s, nine Pacific white-sided dolphins and 17 bottlenose dolphins made up the majority of individuals acquired for the captive cetacean research program as well as four Dall's porpoises and one North Atlantic right whale. The first of these to be given major public attention was Tuffy, a male bottlenose dolphin, who was reported in 1967 by one popular monthly publication to be, like the rest of his species, an intelligent, happy, friendly and playful individual who would 'respond, like children or dogs, to praise and petting' (Stimson, 1967: 178). Unsurprisingly, in this account, the discourse on the military exploitation of the species-specific communication capacities of dolphins minimised the limitations of humans in underwater environments. Human achievements in training and technology were extolled while Tuffy's individual capacities and species-specific abilities were attributed with value only in the context of their deployment for military purposes. Described as an 'efficient diver's helper' during the 1965 Sealab II experiment, Tuffy was trained to take messages and small tools from the surface to the experimental underwater habitat 200 feet below in response to buzzer signals (1967: 69). In preparation for Sealab III, he was expected to take a rescue line to divers stranded at depths of up to 400 feet. The report noted that the navy was 'investigating the idea of teaching marine animals to do useful work' and 'become useful team members' (1967: 68). In this way, the public facing discourse on such endeavours attributed Tuffy and other individual named captive cetaceans with personality and agency, referring to them as 'recruits' and suggesting that they stayed in the program by choice. In this latter regard, two similar accounts, 36 years apart, of individuals being released and returning willingly to captivity were reported. In the first case, Tuffy and a female dolphin called Peg had been locked in an offshore floating pen, which had been opened by means unknown, and 'the two animals had disappeared' (ibid.). The account did not suggest that Tuffy and Peg had escaped captivity. Instead, readers were told that Tuffy had been sighted 30 miles away from the pen the day after his 'disappearance'. Using his call signal to keep him in one location, Tuffy remained with a trainer until a boat arrived. The account then takes on the language of a rescue narrative: 'Then the trainer helped manoeuvre Tuffy into a sling, to be hoisted safely on board' (ibid.). Three days after the escape, Peg was spotted following a small boat. The article quoted the head of the Marine Bioscience Facility at the Naval Missile Center who said, 'our lab people boarded the boat and led Peg the six miles back to Point Mugu, where she swam docilely into the floating pen' (1967: 69).

A later 2003 account of the beluga whales NOC and Muk being freed by animal rights activists bears similarities with the story of Tuffy and Peg: 'NOC eventually swam back into the enclosure. Muk headed down the coast, ending up 20km away. But the training regime paid off, as Muk dutifully followed the Cold Ops boat back to her pen' (Pugliese, 2003). With the program at that time attracting much scrutiny and public criticism following accusations of animal abuse, one of the trainers interviewed for the article reflecting on NOC and Muk's willingness to go back to captivity reasoned, 'Animals that were poorly treated [. . .] wouldn't willingly return to their abusers' (ibid.). By this reckoning, Muk and NOC were complicit in their own captivity while the article, in referring to the success of the 'training regime', acknowledged the power dynamic between the humans and dolphins and framed it as a triumph of human control over cetacean agency. In the cases of Tuffy and Peg, NOC and Muk, those who narrated the official stories of their training and escapes for the purposes of public consumption had to grant enough human-cetacean similitude to stem any ethical concerns balanced by an acknowledgment of species-specific differences to justify their exploitation.

Tuffy was considered to be a first great success for the navy program. One reason for this was due to his being the first cetacean who could work untethered in open water and would return on command. Ideas about the control of human and nonhuman animal minds by military and intelligence agencies captured the public attention following the publication in 1979 of John D. Marks' *The Search for the Manchurian Candidate: The CIA and Mind Control*. In it, Marks wrote about experiments that involved placing electrodes in the brains of some nonhuman animals and cited an official CIA document from October 1961 which stated that 'the feasibility of remote control of activities in several species of animals has been demonstrated' (Marks, 1979: 181). Although the book focused primarily on humans, Marks did note that trained dolphins had been used during the Vietnam War in a 'swimmer nullification' program in which 'government scientists trained dolphins to attack enemy frogmen with huge needles attached to their snouts. The dolphins carried tanks of compressed air which, when jabbed into a deepdiver cause him to pop dead to the surface' (Marks, 1979: 129). A footnote explained that dolphins sent to Vietnam had escaped from their pens, some of whom eventually returned, 'their bodies and fins covered with attack marks made by other dolphins' (129). According to Marks, their escape was 'unheard of behaviour for trained dolphins' (ibid.). The persistence of rumours about weaponised dolphins persisted with international press coverage in the 1980s and 1990s and the publication of other 'insider' accounts which described how dolphins were outfitted with a compressed gas needle device designed to create an embolism when enemy divers were attacked (O'Barry, 2012; Webb, 2014). In 1990, an article in the *New York Times* cited the claims of former Navy dolphin trainers that 'the animals are bring taught to kill enemy divers with nose-mounted guns and explosives' (New York Times, 1990). Although the weaponisation of dolphins and other cetaceans was strenuously denied by the US Navy, animal advocates maintained that the accounts were truthful and this lent public support to a lawsuit filed in 1989 against the Navy's use of 16 dolphins to guard the Trident Nuclear Submarine Base at Puget Sound, Washington. Press reports claimed the Navy to be 'unmoved by critics', but by May 1990 the Navy had agreed to suspend the project and halt plans to take more wild dolphins into the program pending a review (New York Times, 1990).

As public support for the 1989 lawsuit revealed, popular attitudes towards cetaceans had shifted dramatically between the 1950s and 1990, shaped in no small part by their cultural depictions. There was minimal popular interest in whales at the end of the 1950s, but by the 1970s a popular discourse of cetacean intelligence and beauty was established and public pressure was a key driver in the establishment of legislation to protect marine mammals. As D. Graham Burnett argues, whales had been characterised as 'wily' or 'monstrous', and although there were a 'small number of early voices calling for conservation of the world's largest whales, these animals remained in the early 1960s little more than an industrial commodity of dwindling importance' (Burnett, 2012: 521). By the 1990s, surveys showed that a majority of Americans objected to whaling, and one study claimed that 70 per cent of Americans objected to killing whales on moral grounds (Rose et al., 2009: 39; Kellert, 1999). In the 1960s and 1970s popular depictions of whales – those that emerged concurrent with the shift in public attitudes – were inextricably linked to the scientific study and military exploitation of cetacean intelligence and language. Indeed, work on interspecies communication undertaken in the 1960s established a complex relationship between the US entertainment industries and military bioscience and shaped significant popular depictions of cetaceans as minded speaking agents.

Marine mammal displays had been popular attractions since the late nineteenth century, but the popular perception of cetaceans as intelligent beings was not evident until the middle of the following century. The first captive cetaceans to be exhibited were beluga whales. P.T. Barnum began exhibiting belugas in 1861 initially at the Boston Aquarial Gardens and, after the facility closed, at Barnum's American Museum in New York. In November 1861 Barnum wrote to the *New York Times*:

> I leave this monster leviathan to do his own 'spouting', not doubting that the public will embrace the earliest moment (before it is forever too late) to witness the most novel and extraordinary exhibition ever offered them in this City.
>
> (Barnum, 1861)

As Barnum's account suggested, whales were considered an attraction, monstrous and spectacular, but there was no sense in which they were construed as minded agents. In 1938, Marine Studios, later named Marineland, opened in Florida. A popular tourist attraction, Marine Studios billed itself as the first oceanarium and included a bottlenose dolphin exhibition, although it would not have trained dolphins until the 1950s. During that decade, there was a notable increase in the popularity and number of oceanariums, particularly those that exhibited marine mammals. Marineland of the Pacific, on the Palos Verdes Peninsula coast, opened in 1954 and Miami Seaquarium was founded in 1955. Animal Behavior Enterprises, a company established by two former students of B.F. Skinner and specialising in behaviour analysis and operant conditioning, was established in 1947. The founders of the company were contracted by Marine Studios, Marineland of the Pacific and the US Navy to develop marine mammal shows and training programmes.

In 1955, John C. Lilly, a neuroscientist who had previously conducted in vivo electrical brain stimulation on non-anesthetised macaques, was one of eight investigators given access to dolphins at Marine Studios. The aim was to anesthetise the dolphins and undertake cortical mapping on their exposed brains. Claimed to be a 'success', five dolphins were euthanised during the expedition (Burnett, 2012: 568). There were two further visits made by Lilly to Marine Studios which resulted in the deaths of three more dolphins. By the end of the 1950s, Lilly had begun to develop a theory that dolphins were intelligent, used intraspecies language and had attempted interspecies communication. In 1961 he published *Man and Dolphin* followed later by *Communication between Man and Dolphin: The Possibilities of Talking with Other Species* (Lilly, 1978). In the latter, Lilly wrote about the visits to Marine Studios:

> I found that the dolphin had been making airborne sounds that, when slowed down, sounded like human speech. This observation was key to our subsequent work. It established that the dolphins would do anything to convince the humans that they were sentient and capable.
>
> (Lilly, 1978: 19)

Lilly presented findings of these and other studies on sensory deprivation research to representatives from the Office of Naval Research, Air Force and Army in 1959 (Burnett, 2012).

In 1960 and under Lilly's leadership, the Communication Research Institute (CRI) was established, funded by the National Science Foundation, Office of Naval Research and the Department of Defense. Further funding from NASA supported a particular strand of research; 'investigations on the mechanisms of inter- and intra-species communication of intelligent information, emotional status, and basic drives, in an attempt to discover the mechanisms which nature has evolved, and to supplement these mechanisms by technological devices' (Reynolds, cited in Burnett, 2012: 582).[7] In 1961, Leo Szilard published *The Voice of the Dolphins and Other Stories*, a collection of satirical short stories, the longest of which was the tale of dolphin intelligence based loosely although recognisably on Lilly's work following conversation between the two in 1958 (Lilly, 1978: 20). Lilly's

Man and Dolphin, also published in 1961, was a best-seller. Given extensive coverage in *Life* magazine, images of a dolphin called Elvar pictured in a small Plexiglas tank at the Communications Research Institute accompanied an article titled 'He barks and buzzes, he ticks and whistles, BUT CAN THE DOLPHIN LEARN TO TALK?' (Life, 1961: 61). It is perhaps easy to see why Lilly's work and words captured the public imagination. He wrote: 'It is my firm conviction that within the next decade or two human beings will establish vocal communications with another species' (Lilly, 1961: 68). Whales and elephants, Lilly argued, had brains large enough for 'high level mental activity', dolphins talked with one another and had intelligence of a 'high order' and 'might possibly be taught to understand and react to sounds made by man' (Lilly, 1961: 68). His tape collection, Lilly explained, included examples of dolphins imitating his words and, he contended, 'when a dolphin has to obtain satisfaction of his wants from humans he can be forced to communicate vocally with them' (ibid.). In a disturbing (at least for the contemporary viewer) scene from the 1973 film *The Day of the Dolphin*, Lilly's forced communication scenario is played out when two dolphins named Fa and Bea, are forcibly separated, much to the dolphins' distress. Worried that Fa is 'learning his own language' from Bea, Dr Jake Terrell, played by George C. Scott, watches on as Fa circles furiously in the tank and tries to break through the barrier to be reunited with Bea. Eventually Fa gives up and speaks in a haunting childlike voice: 'Fa wants Bea'. Having asked for what he wants in the appropriate (human) language, Terrell allows Fa and Bea to reunite. At the end of the film, Terrell must release Fa and Bea back to the ocean to keep them safe. In a dialogue with Fa, Terrell explains that he and Bea must go, stay away from humans and 'not talk'. The poignancy of the scene hinges on the dolphins literal infantalisation and their display of cutified devotion to Ferrell and his wife as Fa tells them 'Fa loves Pa' and 'Fa loves Ma' and struggles to understand why he and Bea must leave. The scene ends with Terrell/Pa rebuking Fa and ignoring his plaintive cries as he walks away from the water – a re-working of the noble-sacrifice-for-love ending.

The Day of the Dolphin was based on the 1967 novel *Un animal doué de raison*, a science fiction novel by Robert Merle and inspired by Lilly's work. Published in English in 1969 under the title *The Day of the Dolphin*, Roman Polanski was initially interested in making it into a film but eventually dropped the idea. Polanski introduced director Mike Nichols to the book and Nichols decided to develop a script. The film adaptation bore little resemblance to the book, which Nichols regarded as anti-American. He brought writer-actor Buck Henry on to the project to write a script that would eliminate what Nichols regarded as anti-American 'bias' (Nichols in Chase, 1973: 68). In an interview for *New York Magazine* in 1973, Nichols gave a lengthy account of working with dolphins, who he likened to 'temperamental human actors' and after comparing interspecies communication with alien contact, the director closed the interview with the comment:

> the two main dolphins finished their last shot and swam off into the ocean. We had decided that when we were through, we would set them free. But they took care of that; they set themselves free. Only they did their job first.
>
> (Nichols, quoted in Chase, 1973: 70)

Lilly's work on interspecies communication was also fictionalised in the film *Namu, the Killer Whale* (1966), produced by Ivan Tors, who was also responsible for bringing *Flipper* (1963) to the screen. Tors brought Lilly in as a consultant on *Flipper*, and it was during this time that Lilly convinced the producer that he should make a film the featured an orca. Lilly reasoned that an orca 'would probably accept relationships with man in the water' and he claimed that the film was important in that it challenged the received wisdom about 'killer whales' gleaned from the accounts of whaling ships and Scott's Antarctic diaries. Lilly wrote: 'scenes showing a man riding the back of Namu, holding on to the six-foot-high dorsal fin [. . .] were adequate demonstrations that these were gentle compassionate, cooperative animals' (Lilly, 1978: 15–16). Namu was 'played' by an orca named Namu who was captured off the coast of western Canada in 1965. He had become caught in fishing nets in Namu Bay, reportedly trying to help another younger trapped orca escape. The rest of their pod stayed close by the imprisoned whales while the owners of the nets attempted to make a deal to sell them (Whitehead, 1965: 54–55). The younger whale eventually escaped and the remaining older orca, Namu, was bought for $8,000 to be put on display in a Seattle aquarium. In interviews, the aquarium owner claimed that it was possible to talk to 'killer whales'. He proposed that Namu would be important to scientific research and stated, 'They have more brains than porpoises. Killers are the smart-est things that swim. This whale will be very valuable for research projects. We'll tape his vocabulary' (Whitehead, 1965: 55).

Namu was moved to the floating pen at Pier 56 Seattle Marine Aquarium in July 1965. Tors visited the aquarium and made a deal with Griffin to make a film with Namu. The agreement included the condition that Griffin swim with Namu, as the film's financing depended on this being captured on camera. Before filming began, Griffin demonstrated to Tors that he was able to hold Namu's dorsal fin and be pulled through the water. Namu was then moved again to Rich Cove to begin filming for *Namu, the Killer Whale* in early 1966. The cove was blocked at the mouth with Navy surplus anti-torpedo netting and Namu was towed in to the cove in a floating pen. Most of the scenes involving Namu were shot in March 1966. When filming ended, Namu was returned to the Seattle Pier 56 pen where he developed skin rashes, breathing problems and an infection, which led to his death in July 1966. Popular reports stated that his death was accidental, that he 'had become enmeshed in the nets of his pool pen and had drowned' (Coombs, 1966: 12). The autopsy revealed that Namu had colic and delirium caused by a bacte-rial infection aggravated by pollution. Namu was around 17 years of age when he died. It was established after his death that he had been a member of C1 pod and was given the alphanumeric identifier C11. The orca who was assumed to be his mother (C5) died 29 years later, aged 71. In the fictionalised narrative, Namu is held in a cove by a scientist who teaches him to talk. The local villagers initially believe Namu is a dangerous killer, but they eventually change their minds. At the end of the film Namu returns to open water, finally free.

Namu's life after capture and his death were public events played out through the press coverage of the first 'successful' capture of an orca and the wide-spread fascination with interspecies communication. There is no doubt that the popularisation of Lilly's work and his relationship with Ivan Tors that resulted in *Flipper* and *Namu* were important in challenging the popular understanding about dolphins and orcas. However, Lilly's credibility in the scientific community was eroded after the publication of his first popular book, and while he was able to secure funding for further research (later financial support came from Ivan Tors), there was a wider reluctance to pursue 'talking cetacean' research further. Moreover, Lilly's exploitation of dolphins was later subject to public scrutiny when details of his experiments and casual disregard for the wellbeing of the individuals involved in his experimental practices were reported.

Throughout the factual and fictional re-imaginings of animal language there is the ever-present matter of animal agency. Some of these narratives of animal language tell stories about the limitations of humans in aquatic spaces, where the capacities and labour of other animals are exploited and their training is conceived of as a human victory and control over animal agency. Normalised in situations where anthropocentric interests are centralised, agency is constructed as animals being complicit in their own capture, but a more interesting and perhaps authentic imagining of agency is in animals' acts of escape, resistance, their refusal to cooperate with the filming or penning or expectations of trainers. In this sense, agency is managed at sites of anthropomorphism where animal labour and agency are constructed in relation to their meeting the requirements of human expectations and needs. At the same time, narratives that extol the human success over animal agency in extracting the right type of animal labour usually reveal an authentic animal agency in the places where animals are 'difficult', refuse to be constrained and controlled, where they 'fail' human tests and are problematic narrative plot points.

Conclusion: listening

Control of marine mammals in a navy program was briefly imagined in Willian Gibson's short story *Johnny Mnemonic*, where the character of Jones, a cyborg dolphin, technologically enhanced but surplus to military requirements now lives as a 'War Whale' exhibit in a theme park. With articulated body armour and engineered sensor units on his head, Jones 'was more than a dolphin, but from another dolphin's point of view he might have seemed something less' (Gibson, 1986: 23). In military service, Jones had been used to find cyber-mines in the Pacific Ocean. Talking about Jones' heroin addiction, one of the main characters in the story asks, 'I can see them slipping up when he was demobbed, letting him out of the navy with that gear intact, but how does a cybernetic dolphin get wired to smack?' The female protagonist replies, 'The war. [. . .] They all were. The Navy did it. How else you get 'em working for you?' (Gibson, 1986: 26). In the character of Jones, Gibson imagined the meshing together of various different dolphin realities: the military dolphins of the NMMP, the dolphins used in Lilly's interspecies communication and LSD experiments. The nightmare envisioning of Jones, surplus to requirements, a sideshow attraction and bearing the bodily and mental scars of human abuses did not stray far from the collective experiences of some individuals.

Making human-like sounds is a particularly taxing activity for a beluga whale. In NOC's case he had to vary his nasal tract pressure, over-inflate his vestibular sacs and make muscular adjustments to his vibrating sonic lips (Ridgway et al., 2012). Human fascination with language acquisition in other species does not equate to similar levels of exertion on our part. Indeed, the urge to make other species emulate human communication is perhaps better described as anthropocentric indolence. If we were really open to listening to what other species had to say, this should not require them to 'speak in our voice'. The desire for interspecies communication might be strong but this does not mean that we are prepared to engage in conversation with other species. *Animal Envy: A Fable* (2016) imagines what would happen if nonhuman animals were, through a digital translation application, able to speak to humans in an event called the TALKOUT. In what can be conceived of as an apologetic self-reflexive gesture to the novel's own discursive limitations, within the first few pages we find that the nonhuman animal collective 'knew that to make their core message palatable, it would have to be framed in a way that was ingratiating and flattering to humans' self-interest' (Nadar, 2016: 6). Reflecting human strategies for communicating a moderate environmental and welfarist position to a mainstream audience, the narration tells us the nonhuman animals decide that 'They were not going to rely on appeals to justice or fairness, but on showing human animals how useful the animal kingdom can be to humans the more humans understand them' (ibid.). Although the book tries to imagine a space in which nonhuman animals could make their case to humans, all that they are given to say is what is already known. 'Animal scientists and ecologists already know', we are told, 'but it will have more effect coming straight from the source, the animals felt' (ibid.). As Josephine Donovan observes, 'We should not kill, eat, torture or exploit animals because they do not want to be so treated, and we know that. If we listen, we can hear them' (2007: 76). In this sense, *Animal Envy* makes the point that we already have the capability to 'listen'.

The insistence that communication takes place on human terms has been construed as an enhancement to the linguistically diminished worlds of other animals. But, what popular culture has imagined at times is that such anthropocentric bias diminishes the individual subjectivity of others who are consigned to a liminal existence, materially and metaphorically, where they are denied full entry to human and nonhuman animal worlds. What this suggests is that language only serves to tip the scales further in favour of humans and the preferred asymmetries of power: it does not act as a corrective to already unequal speciesist power relations. We might give other animals a voice so that they can say what we want to hear, a move that reveals the anthropocentric bias of anthropomorphism. In this regard, popular culture has taken a key role as the purveyor of interspecies communication narratives and chief mediator of nonhuman animal subjectivity and experience. It is therefore important to consider who speaks and whose interests are being served in these narratives. *Finding Dory* (2016), an animated feature, offers a representation of 'talking animals' that acknowledges similitude and difference and takes a refreshing, if minor, detour away from other zoomorphic depictions. In the film, a beluga whale named Bailey becomes a hero by using his echolocation, a species-specific trait which is used to gain an advantage over the humans. There is thus at least an acknowledgement of species difference and interests and a more satisfying narrative nod to rebalancing the relations of power between humans and other animals.

When humans speak on behalf of other animals, the discursive constraints imposed by anthropocentric self-interest are the ever-present risk. The challenge is not in speaking but in listening. A conversation with other species would include an obligation to respond, and that response would be costly in terms of normative human practices. A conversation with another 'speaking' mind has moral responsibilities. It is indeed strange that humans are so keen to silence nonhuman animals while expending so much energy and resources in attempts to speak to extra-terrestrial nonhumans and break what is known as the Great Silence. The careful guarding of language as a human-specific characteristic locks us into an echo chamber where only the voices of our own anthropocentric concerns can reverberate. While we continue to revere the self-congratulatory cacophony of human articulations as confirmation of species advantage, we expand our self-made *silentium universi* by choosing not to listen to other animals.

Notes

1 These features are commonly cited as a selective group from the longer set of 16 features proposed by Charles Hockett (1968) to define language.
2 'Talking Dogs – A Funny Talking Dog Videos Compilation 2016 (www.youtube.com/watch?v=F_kT_lvqsmA); 287 comments analysed.
3 Such content has been a regular feature of the clip-based reality television programming since the 1990s.
4 This is the number of views between 2008 and up to her death in 2017.
5 From Ceta-Base, a database of captive marine mammals (www.cetabase.org/).
6 The website states that the marine mammals in the program live longer than their counterparts in the wild (NMMF). However, the study cited for the claim refers only to bottlenose dolphins in the Marine Mammal Programme from 2004 to 2014.
7 A representative from Marine Studios assisted in securing the first two dolphins for CRI, but both individuals died shortly after their arrival.

References

Aldrich, B.C. (2018) 'The use of primate "actors" in feature films 1990–2013' in *Anthrozoos*, Vol. 31 (1), pp. 5–21.
Anderson, S. (2006) *Doctor Doolittle's Delusion: Animals and the Uniqueness of Human Language*, Yale University Press, New Haven and London.
Arluke, A. and Sanders, C. R. (1996) *Regarding Animals*, Temple University Press, Philadelphia.
Armbruster, K. (2013) 'What do we want from talking animals? Reflections on literary representations of animal voices and minds' in DeMello, M. (ed) *Speaking for Animals: Animal Autobiographical Writing*, Routledge, New York and London, pp. 17–34.
Barnum, P.T. (1861) Letter in *New York Times*, 22 November, online at www.nytimes.com/1861/11/22/news/brooklyn-murder-trial-newfoundland-telegraph-line-weather-st-johns-nf-markets.html?pagewanted=2

Batt, E. (2012) 'Op-ed: The truth behind NOC, the beluga whale who mimicked human speech' *Digital Journal*, 25 October, online at www.digitaljournal.com/article/335503

BBC News (2012) 'Beluga whale "makes human-like sounds"' *BBC News*, 22 October, online at www.bbc.co.uk/news/science-environment-20026938

Beck, A. M. and Katcher, A. H. (1996) *Between Pets and People: The importance of Animal Companionship*, Purdue University Press, Indiana.

Berwick, R. C. and Chomsky, N. (2016) *Why Only Us: Language and Evolution*, MIT Press, Cambridge, MA.

Burnett, D. G. (2012) *The Sounding of the Whale*, University of Chicago Press, Chicago and London.

Catsoulis, J. (2012) 'Chimps eat, scratch, groom' *The New York Times*, 19 April, online at www.nytimes.com/2012/04/20/movies/chimpanzee-a-disney-film-narrated-by-tim-allen.html

Charles, N. (2016) 'Post-human families? Dog-human relations in the domestic sphere' in *Sociological Research Online*, Vol. 21 (3), p. 8, online at www.socresonline.org.uk/21/3/8.html

Charles, N. and Davies, C.A. (2008) 'Family and other animals: Pets as kin' in *Sociological Research Online*, Vol. 13 (5), online at www.socresonline.org.uk/13/5/4.html

Chase, C. (1973) 'Mike Nichols: Man on the dolphin' in *New York Magazine*, 17 September, pp. 68–70.

Choi, C. (2012) 'Beluga whale named NOC mimics human noises with spot-on imitation (audio)' *Huffington Post*, 22 October, online at www.huffingtonpost.com/2012/10/22/noc-beluga-whale-human-voice-imitation_n_2001993.html

Cole, M. and Stewart, K. (2014) *Our Children and other Animals: The Cultural Construction of Human-Animal Relations in Childhood*, Routledge, New York and London.

Connor, S. (2012) ' "Who told me to get out?": NOC the talking whale learns to imitate human speech in attempt to "reach out" to human captors' *Independent*, 22 October, online at www.independent.co.uk/news/science/who-told-me-to-get-out-noc-the-talking-whale-learns-to-imitate-human-speech-in-attempt-to-reach-out-8221800.html

Coombs, C. (1966) 'On the screens' in *Boys' Life*, September, p. 12.

Crichton, M. (2015) 'In his own words' *The Official Site of Michael Crichton*, online at www.michaelcrichton.com/congo/

Dale, J. P., Goggin, J., Leyda, J., McIntyre, A., and Negra, D. (2017) *The Aesthetics of Cuteness*, Routledge, London and New York.

de Saussure, F. (1915) *Course in General Linguistics*, McGraw-Hill, New York, Toronto and London.

DeMello, M. (2013) 'Introduction' in DeMello, M. (ed) *Speaking for Animals: Animal Autobiographical Writing*, Routledge, New York and London, pp. 1–16.

Disney (2012) 'Disney's use of live animals in entertainment policy', 3 April, online at https://ditm-twdc-us.storage.googleapis.com/Disneys-Use-of-Live-Animals-in-Entertainment-Policy.pdf

DogForums (2012) 'Husky noises' online at https://www.dogforums.com/first-time-dog-owner/116961-husky-noises.html

Donavan, J. (2007) 'Animal rights and feminist theory' in Donovan, J. and Adams, C. J. (eds) *The Feminist Care Tradition in Animal Ethics*, Columbia University Press, New York, pp. 58–86.

Friedman, E. (2013) 'The role of pets in enhanced human well-being: Physical effects' in Robinson, L. (ed) *Waltham Book of Human-Animal Interaction: Benefits and Responsibilities of Pet Ownership*, Pergamon, Oxford, pp. 33–54.

Fudge, E. (2002) *Animal*, Reaktion Books, London.

Gibson, W. (1986) 'Johnny Mnemonic' in *Burning Chrome*, HarperCollins, London, p. 14–36.

Haraway, D. (1989) *Primate Vision: Gender, Race and Nature in the World of Modern Sciences*, Routledge, New York and London.

Hockett, C. F., and Altmann, S. A. (1968) 'A note on design features' in Sebeok, T. (ed) *Animal Communication: Techniques of Study and Results of Research*, Indiana University Press, Bloomington, pp. 61–72.

Jane Goodall Institute (n.d.) 'Letter from Jane Goodall to professionals in the entertainment and advertising industry', online at www.janegoodall.org.uk/15-chimpanzees/chimpanzee-central/27-chimpanzees-in-entertainment

Jenkins, H., Ford, S., and Joshua, G. (2013) *Spreadable Media: Creating Value and Meaning in a Networked Culture*, New York University Press, New York.

Joyce, C. (2012) 'Baby beluga, swim so wild and sing for me' *NPR*, 23 October, online at www.npr.org/2012/10/23/163471645/baby-beluga-swim-so-wild-and-sing-for-me

Kellert, S. R. (1999) *American Perceptions of Marine Mammals and Their Management*, Humane Society of the United States, Washington, DC.

Life (1961) 'Can the dolphin learn to talk?' in *Life*, 28 July, pp. 61–66.

Lilly, J. (1961) 'Importance of being Earnest about dialogues of dolphins' in *Life*, 28 July, p. 68.

Lilly, J. C. (1978) *Communication between Man and Dolphin: The Possibilities of Talking with other Species*, Crown Publishers, New York.

Lumenick, L. (2012) 'Chimpanzee' *New York Post*, 20 April, online at https://nypost.com/2012/04/20/chimpanzee/

Marks, J. D. (1979) *The Search of the Manchurian Candidate: The CIA and Mind Control*, Time Books, New York.

Molloy, C. (2011) *Popular Media and Animals*, Palgrave Macmillan, Basingstoke.

Molloy, C. (2012) 'Being a known animal' in Blake, C., Molloy, C., and Shakespeare, S. (eds) *Beyond Human: From Animality to Transhumanism*, Continuum, London and New York.

Molloy, C. (2013) 'Nature writes the screenplays: Commercial wildlife films and ecological entertainment' in Rust, S., Monani, S., and Sean, C. (eds) *EcoCinema Theory and Practice*, Routledge, London and New York.

Morgan, D. (2001) *Siberian Huskies for Dummies*, Wiley, Indiana.

Nadar, R. (2016) *Animal Envy*, Seven Stories Press, New York.

Nenkov, G. Y. and Scott, M. L. (2014) '"So cute I could eat it up": Priming effects of cute products on indulgent consumption' in *Journal of Consumer Research*, Vol. 41, pp. 326–341.

New York Times (1990) 'Navy suspends a plan to use dolphins as guards' *New York Times*, 23 July, online at www.nytimes.com/1990/07/24/us/navy-suspends-a-plan-to-use-dolphins-as-guards.html

Ngai, S. (2012) *Our Aesthetic Categories*, Harvard University Press, Cambridge, MA and London.

O'Barry, R. (2012) *Behind the Dolphin Smile*, Earth Aware, California.

Page, A. (2017) '"This baby sloth will inspire you to keep going": Capital, labor and the affective power of cute animal videos' in Dale, J. P., Goggin, J., Leyda, J., McIntyre, A., and Negra, D (eds) *The Aesthetics of Cuteness*, Routledge, London and New York, pp. 75–94.

Pierce, J. (2016) *Run, Spot, Run: The Ethics of Keeping Pets*, Chicago University Press, Chicago.

Prigg, M. (2012) 'The whale who learned to talk to the man-imals: NOC could imitate a human's voice and used an underwater microphone to make contact with scientists'.

Mail Online, 22 October, online at www.dailymail.co.uk/sciencetech/article-2221405/
Incredible-audio-reveals-white-whale-trying-make-contact-humans.html

Pugliese (2003) 'Secret weapon' *The Scotsman*, 8 March, online at www.scotsman.com/
lifestyle/secret-weapon-1-599237

Rensberger (1972) 'Invasion by aggressive honeybees is feared' *New York Times*, 22 January, online at www.nytimes.com/1972/01/22/archives/invasion-by-aggressive-honeybees-is-feared.html

Ridgway, S., Carder, D., Michelle, J., and Marl, T. (2012) 'Spontaneous human mimicry by a cetacean' in *Current Biology*, Vol. 22 (20).

Rose, N.A., Parsons, E.C., and Farinato, R. (2009) *The Case Against Marine Mammals in Captivity*, The Humane Society of the United States and the World Society for the Protection of Animals, Washington, DC.

Savage- Rumbaugh, S., Shankur, S., and Taylor, T.J. (2001) *Apes, Language and the Human Mind*, Oxford University Press, New York.

Scale, H. (2012) ' "Talking" whale could imitate human voice' *National Geographic*, 23 October, online at http://news.nationalgeographic.com/news/2012/10/121022-whales-voices-science-animals-humans-marine-mammals/

Serpell, J. A. (2003) 'Anthropomorphism and anthropomorphic selection – Beyond the "cute response" ' in *Society and Animals*, Vol. 11 (1), pp. 83–100.

Spong, P. (2011) 'Communication' in Brakes, P. and Simmonds, P. (eds) *Whales and Dolphins: Cognition, Culture, Conservation and Human Perceptions*, Earthscan, London and Washington, DC.

Stimson, T. (1967) 'Tuffy, the Navy's deep sea lifeguard' in *Popular Mechanics*, July, pp. 66–69, 178.

Webb, B. (2014) *The Red Circle*, St. Martin's Press, New York.

Whitehead, E. (1965) 'Conversation-starved killer in a salmon net' in *Sports Illustrated*, 12 July, pp. 54–59.

Williams, J. (2012) 'Rotten tomatoes Disneynature chimpanzee reviews', online at www.rottentomatoes.com/m/disneynature_chimpanzee/reviews/?type=top_critics

Wittgenstein, L. [1921] (2001) *Tractatus logico-philosophicus*, Routledge, London.

6   When animals think

Introduction

In June 2014, a *Guardian* news article titled 'Five insights challenging science's unshakable "truths"' listed a group of 'unexpected' scientific findings. At number five, the article declared in somewhat lacklustre terms that 'human beings are nothing special' (Brooks, 2014). Exclusivity, even at a genetic level was, the article noted, minimal and all those characteristics that we thought lined the chasm of difference between humans and other species – abstract reasoning, personality, morality, culture, feelings – are evident in crows, chimps, dolphins, rats, spiders, even cockroaches who, the article admitted, 'have feelings, too, it turns out' (Brooks, 2014). The article concluded with the question: 'The next step may be more far-reaching: how comfortable would we be, for instance, eating a lobster that we knew was terrified by its capture?' It perhaps speaks to some middle-class concern that a lobster should be the example the author selects – why not a cow, a pig or a chicken or one of the species already mentioned in the article who are also routinely killed by humans? But if we set aside the apparent randomness of the species, it is significant that, by posing the problem, the author takes the inevitable step from animal minds to animal ethics. In this concluding chapter I draw together the main themes of the book with a focus on the sites where some animals are constructed as minded and mindless and explore how animal minds and the capacity to suffer have been imagined in popular culture.

Animal minds

If nonhuman animals have minds, and if those minds are in some respect or other like ours, what then are our obligations towards those species? These are not new questions and with the issue of continuity between human and animal minds on the public agenda since Darwin's time, it is perhaps salient to ask why this remains such a controversial question. Press reports on the scientific evidence of animal-human comparability might fascinate audiences but what those studies mean in terms of treatment, rights or welfare implications remain caught up in endless rounds of counter arguments, dismissals, accusations of anthropomorphism and a reinvestment in speciesist claims to human privilege. Conflicting views held about animal minds and whether or not species other than humans have conscious experiences are considered obvious to those on either side of the scientific argument. At the root of the confusion, consciousness remains notoriously difficult to define and even where there is agreement about what consciousness is, the 'problem' of animal consciousness continues to be caught up in the cavernous gap between two camps of thought. Simply put, there exists one argument that observable behaviours are indicators of conscious thoughts and feelings while an opposing perspective would have it that consciousness remains unobservable and inaccessible so behaviours cannot be taken as evidence of animal thinking. While the former position might then argue that cognitive capacities confer moral significance and therefore some degree of rights or interests, the latter argument may conclude that without proof of consciousness nonhuman animals are merely a bundle of behaviours that do not require any special treatment or moral concern.

Some argue for the separation of animal consciousness from the issue of animal welfare (see, for instance, Dawkins, 2012). This view proposes that consciousness only confuses the question of how humans should treat animals because it is erroneously anthropomorphic to assume that animals have mental experiences like humans. For proponents of this argument, welfare should be concerned instead not with the issue of animal minds but general animal wellbeing which can be understood and measured through the observation of behaviours. But in this case, moral concern for animals does not go beyond that which benefits humans, for instance in the form of welfare measures that maintain the 'quality' or the health of the animals who will be slaughtered and consumed. However, when the link between proof of consciousness and moral status is centralised, human cognition continues to be used as the benchmark for moral significance and animals will always occupy an inferior position. By insisting on cognitive similitude between humans and other species as a marker for moral significance, humans are able to maintain their sense of entitlement to exploit other animals. When human cognition becomes the ultimate measure for conferment of rights, this can be conceived of as a form of anthropomorphism which only serves to further anthropocentric views about human privilege in the world. To this there is a counterargument: that it is appropriate that humans have species-specific cognitive capacities *and* moral responsibility while other animals have rights not to be killed, tortured or otherwise exploited by humans *and* to possess their own species-specific ways of relating to the world. This position is not incompatible with anthropomorphism as a means by which we can establish empathetic connections with other animals. In this regard, when 'human' is the benchmark, the problems of anthropocentric pejorative anthropomorphism arise. Using the experience of being a sentient embodied animal as a means to understand the experiences of others, anthropomorphism can be a meaningful part of pursuing pragmatic empathetic connections.

Popular culture, in its various modalities, plays a role in both questioning and setting the contours of public debate on nonhuman animal minds. In terms of appeal, stories about animal consciousness, human-animal cognitive comparability and animal intelligence continue to occupy the public imagination and inform the wider conversation about the interests of animals other than humans. Indeed, animal mind and the implications of consciousness in other species have proven particularly popular across factual and fictional genres, albeit with notable shifts in the way in which such stories are told. In the same year that the *Guardian* article on science's unexpected findings was published, the BBC series *Inside the Animal Mind* and the PBS *Nova* series *Inside Animal Minds* were broadcast. Using the same footage and a similar three-part episode structure dealing with the themes of the senses, problem-solving and the relationship between animal intelligence and social groups, these two television programmes attempted to imagine, with reference to current scientific thinking, what it is like 'to be' a nonhuman animal. Press coverage of the Nonhuman Rights Project, an organisation focused on petitioning to extend legal rights to species other than humans based on scientific evidence of complex cognitive ability, filed its first lawsuits in December 2013. By 2014, the documentary *Blackfish* (2013), which drew attention to the mental suffering of captive orcas, was reported to have influenced public attitudes to such an extent that the press were referring to its impact as 'the Blackfish effect'. News stories about the intelligence of chimpanzees, crows, dogs, cats, elephants and goats were in abundance in 2014, which also saw the release of *Dawn of the Planet of the Apes*, a sequel to the reboot *Rise of the Planet of the Apes* (2011) and a continuation of the science fiction narrative about super-intelligent apes.

In 2011, *Rise of the Planet of the Apes* had been an unexpected success. The rebooted origin story took the well-known franchise in a wholly different direction to all previous iterations with an overt animal advocacy message running through a story told from the ape point of view. Directed by Rupert Wyatt and written by Rick Jaffa and Amanda Silver, *Rise* took in excess of $480 million worldwide and was the number one ranked film during its domestic opening weekend in August 2011. Following a preview at the Sundance Film Festival in January 2011 and one month before *Rise*, the British documentary *Project Nim*, directed by Jim Marsh, had its theatrical release. Taking in a mere $400,000 at the US box office, *Project Nim*, a film about a chimpanzee named Nim Chimpsky and the Project Nim language experiment at Columbia University, could not claim to have the same initial public reach as *Rise*. Nonetheless, it is significant that two films, one a commercial science fiction blockbuster, the other a low-budget documentary, released within a month of each other, dealt with issue of nonhuman primate minds and experience and raised questions about the ethics of using apes for experimental purposes. Reviews were quick to point out that the films shared common themes and that it was notable that both were more sympathetic to the nonhuman animals than to the human characters. The critical response to *Project Nim* was overwhelmingly positive, with reviewers remarking that 'The chimp emerges from this experience as a more admirable creature than many of its humans' (Ebert, 2011), 'The chimp comes out of it well. *Homo sapiens*, of course, is found wanting' (Bradshaw, 2011) and that Nim was 'a helpless innocent compared with his protectors and tormentors' (Scott, 2011). In 2014, zoos were also thrust into the global spotlight when the killing of a healthy young giraffe named Marius by Copenhagen Zoo made headlines and the public was made aware of the routine practice of killing 'surplus' captive animals in European and US institutions. Many of these stories made links to the psychological suffering of animals in captivity. The crowdfunded documentary *Zoochosis* was released in May 2014 and other footage of 'zoochosis' – abnormal or stereotypic behaviours – circulated on social media and provided further visual confirmation for the public that captive animals experience extreme mental distress. What emerged from the popular discourse on animal mentality is that where there are minds, there is the potential for suffering.

Mindlessness

In mainstream commercial fiction films, the presence of a nonhuman animal mind with the capacity to think, strategise, plan and remember emerges as a somewhat different prospect to that of the mindless killing machine, a one-dimensional representation of a species that emphasises to the level of gross distortion a single type of behaviour, epitomised by *Jaws* (1975) and the film's various imitators that followed. This is not to say that monstrous nonhuman animals or thematic anxieties about human exceptionalism disappeared from commercial films. After *Jaws*, sharks remain a popular 'monster' with films such as *Sharknado* (2013) and subsequent movies in the franchise exploiting the mindless killing machine trope as a comedy horror parody of previous revenge of nature films. A representation of the calculating relentless evil shark, capable of matching human strategies for escape, features in the survival thriller *The Shallows* (2016) while independent film company The Asylum, responsible for the Sharknado franchise, produced and distributed the Mega Shark series of films (*Mega Shark Versus Giant Octopus* (2009), *Mega Shark Versus Crocosaurus* (2010), *Mega Shark Versus Mecha Shark* (2014), *Mega Shark Versus Kolossus* (2014)), the multi-headed *Shark Attack* series (*2-Headed Shark Attack* (2012), *3-Headed Shark Attack* (2015), *5-Headed Shark Attack* (2017)), *Shark Week* (2012), *Empire of the Sharks* (2017), *Planet of the Sharks* (2016) and *Ice Sharks* (2016). The Discovery Channel's Shark Week, an event that began in 1988 and timed to coincide each subsequent year with 'beach time', met with increasing criticism for chasing ratings, airing docufiction programmes and perpetuating 'fear and misunderstanding' about 'the most misunderstood animals on our planet' (Cohen, 2014). The race to gain audience viewing figures and capitalise on the lucrative advertising slots that event television such as Shark Week could command resulted, critics argued, in sensationalistic programming that misled audiences and eroded the conservation discourse that the channel aimed to promote. Such criticism is not new but draws attention again to the problematic situation over nonhuman animal representations in the commercial media marketplace. It is perhaps easy to ignore or dismiss the Asylum made-for-TV movies and Discovery Channel docufiction as schlocky, irreverent and having little traction in terms of steering public understanding about sharks. However, such commercial ventures make economic sense and attract significant audiences. In terms of their reception, the Discovery docufictions were met with audience confusion over their veracity. The 'bad' computer-generated imagery

(CGI) in Sharknado films was touted as part of their appeal, with poorly rendered 'sharks' acting as little more than a simple 'threat to life' that has been given bodily form. As a parody of other revenge of nature and disaster movies, the representations of sharks are not meant to be taken seriously. They are little more than CGI teeth and flailing bodies that are kicked, punched and chainsawed in half, or that crush celebrities in cameo roles. Such ironic depictions were once argued to be a postmodern indicator of our sophisticated media literacy, able to reflexively give a mirror to our prejudices and highlight social hypocrisy. The problem however is that in continually perpetuating the trope of shark as monster / shark as threat, the irony of such depictions is eroded by its repetition and cynical detachment, the result of which is an unchallenged and unthinking normalisation of a shark stereotype.

A review of Shark Week in 2014 argued 'it's not new; it's old fashioned anthropomorphism: Sharks are the ultimate antiheroes – warriors, loners and complex creatures' (Doyle, 2014). In defence of its sensationalist approach, the Discovery Channel group president claimed that popular culture 'has evolved to kind of appreciate the fear factor of sharks' and that Shark Week programming reflected 'Americans' appetite to be absolutely challenged on fear levels and fantasy levels and mystery levels' (O'Neill in Cohen, 2014). Although Discovery made commitments in 2015 to move away from sensationalist programming, in 2017 Shark Week opened with the heavily promoted *Phelps Versus Shark: A Battle for Ocean Supremacy* (Discovery, 2017); a race between the swimmer Michael Phelps and a Great White shark. Audiences expressed disappointment on social media when the 'race' aired and viewers realised that the shark was computer generated. Primarily, the stunt aimed to foster the notion that humans are demonstrably superior, equal to, or at the very least close in accomplishments to other nonhuman animals. Prior to the Discovery race, promotion for the event underscored the ridiculous anthropocentric egoism that informed the premise of the stunt: 'the odds are stacked against Phelps. Great Whites are the fearsome predators of the sea that can swim up to 10 times faster than humans. But this is no ordinary human that is being tossed into the water' (Snierson, 2017). A discourse of the 'superhuman' (usually male) who can single-handedly survive, control or tame 'nature' found success in twenty-first-century 'man-versus-nature' nonfiction programming (*Extreme Survival* (1999–2002); *Man vs. Beast* (2003); *Man vs. Wild* (2006–2011); *Man, Woman, Wild* (2010–2012)). Billed as survival television, these gendered performances of human superiority act as a reminder that corrosive anthropocentric fantasies of the 'mastery over nature' narrative continued to have a grip on the public consciousness. 'Man vs. animal' races, used primarily for marketing and promotional reasons (Devin Hester and Chris Johnson vs. a cheetah in 2013 for Nat Geo Wild; Usain Bolt vs. a cheetah in 2012 for Nat Geo Wild; Jamie Baulch vs. a horse in 2010 at Kempton Park racetrack; Dennis Northcutt vs. an ostrich in 2009) all, including Phelps and the shark, resulted in the animal 'winning the race'. Nonetheless, borrowing from the man versus nature discourse and creating the racing stunts in the first place continued to promote the notion of contest between humans and other animals, dressed up as 'sport' with the promise of a survival-style spectacle should the stunt go wrong. Coverage of the television event extolled the brilliance of the trainers who prepared the live nonhuman animals for the event, the human scientists who measured, observed and recorded the event, and the proficiency of human technology in the form of CGI that allowed for the comparisons to be made. Across the piece, animals functioned merely as props in the spectacle of human endeavour.

Nonhuman animals with minds, intention and motive are effective 'monsters' but the attribution of 'humanlike' qualities makes it easier for audiences to identify with them or at least have some insight into the reasons for their actions. Mindless animals, on the other hand, operate as narrative functions given bodily form, simplified unrelenting mechanistic behaviours that 'require of us no more than a belief in the immediate reality of their threat' (Tudor, 1989: 115). Monstrous animals with minds were common in 1970s revenge of nature films, usually appearing as packs or infestations with collective intelligence in the first half of the decade and then as individual animals after 1975 and the success of *Jaws* (Molloy, 2011: 157). Maurice Yacowar observes that 'animal-attack films provide a frightening reversal of the chain of being, attributing will, mind, and collective power to creatures usually considered to be safely without these qualities' (Yacowar, 2003: 277–278). The revenge was usually motivated by environmental pollution or genetic engineering and a common narrative strategy had the animals hunt down and kill the human polluters or scientists responsible for their crimes against nature. These films reflected wider social anxieties about pollution, an issue which had moved onto the mainstream US political agenda in the 1970s, and genetic engineering following press reports about the lack of safeguards and the risks posed by new gene transfer technologies in the UK and US.

Through the attribution of minds and the humanised motive of revenge, audiences could understand and perhaps even sympathise with the monstrous animal's purpose and support the eventual demise of an evil scientist or corporate villain (Molloy, 2011: 146). Nonetheless, the genre constraints of revenge of nature films in the 1970s and 1980s limited the audience's sympathies for animal characters and the monsters usually died by the end of the film. Moreover, concerns expressed in the films were very much anthropocentric and reflected anxieties about what environmental destruction or genetic engineering meant for humans. Nature, in animal form, could be vicious and unrelenting and revenge of nature films tended to maintain a focus on human survival. By way of a contrast, the 2010 film *Furry Vengeance* is a comedy that borrows the revenge of nature tropes, depicting a collective of intelligent species in a forest reserve who take revenge on a real estate developer that plans to turn their forest into a residential development. A family-oriented film, it fared poorly at the box office, generating a little over $36 million worldwide against a production cost of $35 million. By the end of *Furry Vengeance*, the animals have won, the forest is a protected nature preserve and the developer has become a park ranger. Aimed at a wholly different market to the later *Sharknado* franchise but still a revenge-of-nature parody, *Furry Vengeance* reflected changes to the way in which animals and their concerns were being represented by Hollywood. This shift was noticeable during the 1990s with the production and distribution of commercial films with an environmental or animal advocacy message that included *FernGully: The Last Rainforest* (1992), *Once Upon a Forest* (1993), *Free Willy* (1993), *Babe* (1995) and *Buddy* (1997) and the opening, in 1995, of *Circle of Life: An Environmental Fable*, an edutainment film attraction, at Epcot in Disney World, Florida. While these films constituted the emergence of a trend that gathered momentum in the next two decades, they did not indicate a wholesale shift away from the animal monsters of the revenge of nature films from the 1970s and 1980s. Indeed, films such as *Arachnophobia* (1990), *Deep Blue Sea* (1999), *Man's Best Friend* (1993) and *Jurassic Park* (1993) continued to reflect anxieties about the control of nature and the monstrous outcomes of human interference with nature.

Animal minds

During the 1990s public outrage over the treatment of Keiko, the male orca who had appeared as Willy in the film *Free Willy* (1993), and the campaign to release him back to the wild demonstrated the extent to which a fiction film could mobilise public support for animal advocacy. The film depicted the eponymous Willy as a minded being with emotions, interests and language capacities, and in many regards he was an anthropomorphic depiction that was all too familiar to the genre. The film attempted to meld together the affective appeals of love, loss and family, typical of a family tear-jerker movie, with a message about orca exploitation and captivity. Critical reception of the film was disparaging of its sentimentality, with reviewers branding it as 'schmaltzy', 'blatantly manipulative' and 'like Lassie with whales. Only sillier' (Rottentomatoes, n.d.). One critic excused the overly emotional appeals of the film in praise of its 'family appeal' while others opted to comment on its formulaic narrative describing it as 'a sweet-natured-boy-and-his-orca tale for the very young and possibly for their parents or, at least for those graying fans of "Flipper" who are ready to graduate from dolphins to cuddly 7,000-pound killer whales', and 'an unwieldy adaptation of the sturdy old formula about a boy and his dog' (Ebert, 1993; Canby, 1993; Ebert, 1993). Amongst those who expressed disdain for the film there was approval for scenes that combined footage of the real whale, Keiko, and animatronic whales. These comments were not however made in relation to welfare concerns but because the scenes were deemed 'convincing' and technologically smart (Klady, 1993). As one critic noted, 'it's impossible to tell the real and artificial whale apart', and while the review also made the point that 'whales are not as charismatic as dogs, not as easily trained, and cannot be hugged', it conceded that 'the film also does what it can to give Willy a personality' (Ebert, 1993). Commenting further on the comparison between cinematic dogs and orcas, Roger Ebert claimed that the film went beyond the bounds of believability in constructing the orca character:

> it goes a little too far; by the end of the movie, Willy spontaneously figures out how to nod his head for 'yes' and shake it from side to side for 'no', skills that I suspect a real orca would have little interest in developing.

> (ibid.)

Even the more favourable reviews regarded it as manipulative, and most chose to make no comment about or give little attention to the film's commentary on orca captivity.

In one important scene, the trainer (played by Lori Petty) explains to the boy Jesse (played by Jason James Richter) that 'killer whales like Willy live in families: pods. Some of them never leave their moms'. Petty's character goes on to tell Jesse that 'their social structure's important to them' and that 'some of them stay together forever'. While the simplistic explanation functioned to help the audience identify with Willy's plight, make sense of the moment in the film when Willy looks towards the ocean in search of his family, root out the parallel between boy and orca (Jesse is separated from his mother and in a foster home) and reinforce the film's overarching themes of family and loss, Petty's dialogue in this scene is also factually correct. And, for any viewer paying close attention, the trainer's assertion that 'his fin is flopped over. That happens in captivity', and while nobody is sure why it happens, 'maybe they need more room to swim', went beyond the anthropomorphic emotional appeals of parental loss to point out a species-specific consequence of captivity that affected individual whales, in this case both the onscreen fictional Willy and the real Keiko. Moreover the scene makes clear the reason for Willy's captivity is, Petty's character states, that the park's owner 'treats animals like a commodity' before ending with a simplified explanation of how to train an orca. Although the scene is important in terms of establishing character motivations that develop the main plotline which ultimately concludes with Willy's escape, it also gave voice to the interests of real captive orcas exploited by the entertainment industries. Although guilty of the same practice, this was an intentional move on the part of the film's producers who planned that the film would give greater visibility to campaigns to save whales. Richard Donner, the executive producer, was known for his animal rights and pro-choice views and had included animal rights and anti-fur posters, stickers and slogans in scenes in *Lethal Weapon 3* (1992), the third film he directed in the *Lethal Weapon* franchise. In this regard, the dialogue between Petty and Richter was another of Donner's hidden in plain sight animal rights statements. However, *Free Willy*'s message was disregarded by a majority of critics at the time who instead focused on the sentimentality of a formulaic boy-who-loves-an-animal narrative and found orcas to be a poor replacement for the more familiar canine counterpart of similar stories. If the critics saw little reason to be troubled by the film's message the same was not true for SeaWorld. Before the film went into production, SeaWorld, which at the time reportedly held 21 of the 23 captive orcas in the United States, refused the request for one of 'their' whales to star in the movie (Orlean, 2002: 57). This resulted in the producers approaching the Reino Aventura amusement park in Mexico City, where Keiko had been held since 1985.

Despite the mixed critical reviews, the film was a box office success and the campaign to release Keiko back to open waters near Iceland, where he had been captured in 1979, demonstrated that the affective resonances of the film were strong enough to mobilise widespread public support. At the end of the film, a message from the producers accompanied a telephone number 1-800-4-WHALES and encouraged audiences to call if they were interested in saving the whales. The number belonged to the environmental group Earth Island Institute. What had not been anticipated by Warner Bros. or Earth Island Institute was that people would call to ask about the whale who had appeared in the film. The unforeseen public pressure led to a shift in the focus of the film's associated campaign from save the whales to save Keiko, which acquired support from the studio, Warner Bros., and from Michael Jackson, who had written and recorded "Will You Be There" (1993), used as the film's theme song. In 1994 the Keiko Project was created by the Free Willy Keiko Foundation and the Humane Society of the United States (HSUS). The aims of the project were to rescue, rehabilitate and release Keiko back to open waters. Funds to support the project came from the contributions of more than a million members of the public, donations from Warner Bros., New Regency, the HSUS and the Craig and Susan McCaw Foundation. Despite ongoing financial support, expectations that the reality of Keiko's life would mirror that of Willy's in the film were confounded. At the park in New Mexico, Keiko was reported to be suffering with a virus and according to a later account in the *New Yorker* 'possibly even dying' (2002: 57). Keiko's release proved controversial, and following his death in 2003 the market for the Keiko/*Free Willy* story remained buoyant; differing accounts of how human intervention had impacted on his life and death were published in book form (*Killing Keiko: The True Story of Free Willy's Return to the Wild* (2014); *Freeing Keiko: The Journey of a Killer Whale from Free Willy to the Wild* (2005)) and released as a long-form documentary, *Keiko: The Untold Story of the Star of Free Willy* (2016). In this regard, however, the later currency of the Keiko story was in large part due to the public debate on orca captivity that resulted from the documentary *Blackfish* in 2013.

In one sense, *Free Willy* was a forerunner of later issue-driven films that use a film as the centrepiece of a campaign, although later campaigns would by and large come to use, although not exclusively, feature-length documentaries as their focus. *Free Willy* used the narrative strategies and affective appeals typical for a formulaic family film, creating parallels between the whale and main human characters in terms of personality (both are initially rebellious) and background (both have been separated from their mothers) and constructed Willy as a minded agent who experiences loss, isolation, joy and sadness. Combined with an anti-captivity message the film did, at times, go beyond merely humanising Willy by acknowledging his species-specific interests although the constraints of a commercial fiction feature film for children precluded it from going too far with a message. In this sense, there is a line to walk in fiction films between engaging the audience with a message and being so heavy-handed with that message it is perceived to be overbearing or 'preachy'. Nonetheless, there was little to defend the film against the accusation of overt anthropomorphism that relied on human fantasies of family, freedom from captivity, and human-nonhuman animal bonds with a barely noticeable commentary about real orca interests.

Public reviews of *Free Willy* have reflected a reasonably consistent reception of the film's core message of anti-captivity and a high level of engagement with the emotional appeals of the film, particularly the friendship between the two main characters Jesse and Willy.[1] Allied with this, the real names of the two main characters – Keiko and Jason James Richter – were mentioned equally.[2] Where the issue of audience for the film arose, the majority of reviewers, whether giving the film a positive or negative rating, stated that *Free Willy* was a film for children. Nostalgia for childhood and the 1990s framed the affective power of the film's message in public reviews that recalled the film's impact at the time of its original release and then on re-watching it as an adult. There are pleasures associated with repeat film viewing: identification with a specific character; familiarity with the affective impact of the narrative; and, recalling moments from personal history associated with the film or some aspect of its context (see Molloy, 2011; Klinger, 2006). In the case of *Free Willy*, the anti-captivity message was also referred to as an animal rights and conservation message. In those reviews where the film was re-watched in adulthood and the 'message' mentioned, it was recalled in nostalgic terms as a positive aspect of the reviewer's childhood. In some cases, reviewers noted that the film had had a lasting impact on them. When mentioned, the impact was related to a fascination with or a 'love' for orcas, or in relation to the human-orca friendship bond.

There is no doubt that *Free Willy* successfully engaged audiences and was able to act as a mobilising force for action but this should not occlude wider issues about the use of live cetaceans in theme parks and as tourist attractions. For instance, after *Free Willy*, SeaWorld prospered with increases in visitor numbers from 4.9 million in 1997 to 5.6 million in 2005 (Clave, 2007: 127). The public fascination with whales, particularly orcas and dolphins, also grew rapidly and resulted in the expansion of tourism activities specialising in whale-watching and swimming with dolphins. By 2009, whale-watching alone had generated more than $2.1 billion in total revenues across 120 countries and territories with more than 13 million people reported to take a whale-watching trip each year (WDC, 2016; O'Connor et al., 2009). Those involved in advocacy and conservation have argued that responsible whale-watching provides economic benefits that can out-weigh those gained from whale-hunting. As it is difficult for whale-hunting and whale-watching to exist together, advocacy groups have supported the promo-tion of whale-watching but caution that without careful management the activities have deleterious impacts on individuals and groups that include long-term effects on health, habitat shifts, reduced reproductive success, and serious and fatal inju-ries (WDC, 2016). Swimming with whales and dolphins, another lucrative tourist activity, is widely criticised by conservation groups as being highly intrusive and stressful for marine mammals and having long-term impacts on their life pro-cesses as well as being a high-risk activity for humans. The economic argument about whales reminds us that one of the claims made in *Free Willy* – that orcas are commodities – remains true.

Free Willy was part of the shift in Hollywood in the 1990s towards eco-narratives that were aimed at the family market. Without Keiko and using only animatronics, *Free Willy 2: The Adventure Home* (1995) had an environmental message about sea pollution, while *Free Willy 3: The Rescue* (1997) took an anti-whaling stance. A fourth film, *Free Willy: Escape from Pirate's Cove* (2010) rebooted the fran-chise and publicity for the film made clear that no live orcas were used, although in promotional interviews with the cast, children were reassured that they would still be able to see scenes with live penguins and a giraffe (Hogan, 2010). The choice to cast Bindi Irwin, the daughter of the television personality Steve Irwin (known as 'The Crocodile Hunter'), was an attempt to increase the conservationist credentials of the film, the theme of which was orca rescue and rehabilitation. The film did not have a theatrical release in cinemas and was instead released on DVD and Blu-ray under the Warner Premiere label, the division of Warner Bros. that, until 2012, distributed original direct-to-video films focusing on follow-up films that were not expected to be profitable from a cinema release. With the third film in the franchise proving to be a flop at the box office, the decision to go straight to video with the reboot made sense. However, what is interesting about the failing fortunes of the franchise is that, despite the message of the first film, it was suggested that one of the main reasons for the subsequent failures was that the whale was not real. One critic, who had favourable things to say about the first film, wrote: 'unfortunately, this time, Keiko was not used; apparently, she was trapped in a Mexican amusement park with no one to set her free. So Willy was invented anew by special effects experts' (Wilmington, 1997). Declaring the effects 'ingenious', the critic argued that the 'difference between a real mammal and the computer and animatronics effects (and occasional animal shots)' made it difficult 'to feel sympathy when you know – instinctively or otherwise – that the characters are rescuing a computer image or communing with a graphic effect, or hugging and slobbering over an animatronics illusion' (ibid.). The criticism of the animatronic whale draws an interesting parallel with the viewer complaints five years later about the CGI shark used for the Phelps race stunt and reflects a worrying demand for the spectacle of real nonhuman animals. Some critical reviews were positive about the rebooted Willy narrative, because of the film's strong anti-whaling storyline and the use of animatronics was praised with one critic commenting on the considerable 'emotional connection to the audience' (Klady, 1997) while another commended the 'sparkling nature cinematography' (Ebert, 1997). There was recognition that in using special effects, *Free Willy 3* did not have the burden of hypocrisy undermining its message that the first film had to acknowledge by using Keiko. In *Film Journal International*, a reviewer praised the representation of whales and concluded that 'in a year of movie animals shooting basketballs and solving crimes and playing chess, it's pretty refreshing to have a film that treats its human audience with as much respect as its less-evolved subjects' (Luty, 2004). While there was a question mark raised over the extent to which animatronics or CGI could serve the same emotional appeals as live action, other reviews of the third film in the *Free Willy* franchise suggested that it was both possible and preferable to create narratives, even for the family audience, that respected species difference and did not resort to tired signifiers of interspecies communication.

In 2013 and with the release of *Blackfish* the terms of the public debate about marine mammals shifted again. *Free Willy* might have drawn public attention to the issue of orca captivity but organisations such as SeaWorld remained the respectable face of cetacean confinement. *Blackfish* premiered at the Sundance

Film Festival in January 2013 and was broadcast on CNN in October of the same year. Prior to the film's theatrical release in July 2013, SeaWorld shares were valued at $39.65 and by September 2016 the value had sunk to $11.77. In December 2013, eight musical acts withdrew from appearances scheduled for the 2014 SeaWorld Orlando's Bands, Brew and BBQ concert, and reports of a school trip being cancelled as a result of the film received widespread media attention (Mendoza, 2013). After the Sundance premiere, entertainment news reported that representatives from Pixar had viewed *Blackfish* prior to its release and made substantial changes to the script for *Finding Dory* (2016), the sequel to the block-buster *Finding Nemo* (2003). The aquatic park rehabilitation facility had origi-nally been modelled on SeaWorld, one report noted, but 'after seeing Blackfish they retooled the film so that the sea creatures now have the choice to leave that marine park' (Kaufman, 2013). News stories of the 'Blackfish effect' circulated after sharp decreases in SeaWorld attendance were reported in early 2014. In March of the same year, California Assemblyman Richard Bloom introduced the Orca Welfare and Safety Act, which aimed ultimately to phase out orca captiv-ity in California. Legislation introduced by Senator Greg Ball in 2014 sought to prohibit the possession and harbouring of killer whales in aquariums and sea parks as an amendment to the environmental conservation law. SeaWorld began an aggressive rebranding project in 2015 following replacement of the company's CEO and reports that the corporation had experienced an 84 per cent drop in prof-its. The marketing focused on SeaWorld's rescue and rehabilitation programme, the company created an online media campaign 'Ask SeaWorld', launched the 'Blue World Project' to construct larger tanks for orcas and committed to ending its 'Shamu Show' in the San Diego park and replace it with a conservation-based show in 2017. In January 2017, SeaWorld was reported to be in 'damage con-trol mode', a situation exacerbated by Tilikum's death and the introduction of the Orca Responsibility and Care Advancement Bill by US Congressman Adam Schiff in the same month (Rowley and Molloy, 2017). By the end of March 2017, visitor numbers throughout the SeaWorld network of parks had dropped by a fur-ther 14.9 per cent compared with numbers for the same period the previous year (Jones, 2017). SeaWorld also faced a lawsuit brought by shareholders that alleged the company had misled investors about the impact of *Blackfish* on attendance and that it had failed to disclose

> that it had improperly cared for and mistreated its Orca population which adversely impacted trainer and audience safety, that it continued to feature and breed an Orca that had killed and injured numerous trainers, and that consequently created material uncertainties and risks.
>
> (Shareholders Foundation, 2017)

Controversy over *Blackfish* reignited interest in Keiko's story, particularly over questions about the feasibility of releasing captive orcas into open waters. If Tili-kum inherited some of the public feeling about orcas that Keiko's story had initi-ated, released 20 years apart, *Blackfish* was a new generation of orca story.

In terms of its narrative, *Blackfish* subverted the previous tropes of minded and mindless creature from the sea. What was different about the film was that at its core, this was a story of an orca who had killed humans, but in this account the 'killer whale' was the victim. Tilikum was presented through the narrative as a sensitive individual whose mental suffering was a consequence of his captivity that ultimately led to three human deaths. Although not depicted as the mindless killing machine favoured by shark narratives, nor as a cutified Willy-like character, the narrative did 'humanize' Tilikum's experience. Critical reviews praised the film and acknowledged widely that the core message that orcas are intelligent and social mammals who suffer in captivity was compelling and persuasive. In defence of their position, SeaWorld argued that the film was 'propaganda, not a documentary', offered an eight point rebuttal to the film's claims and '69 Reasons Why You Shouldn't Believe *Blackfish*' (SeaWorld, 2017).

Rather than mounting a public rebuttal, SeaWorld had initially contacted film critics directly to outline what it considered to be 'egregious and untrue allegations' and referred to the film as 'shamefully dishonest, deliberately misleading, and scientifically inaccurate' (Batt, 2013). One point of the communication raised objections to the film's anthropomorphism claiming that the implication that killer whales are bullied in zoos or parks was erroneous and that 'the word "bullying" is meaningless when applied to the behavior of an animal like a killer whale' (ibid.). On its website, SeaWorld argued that the film 'conveys falsehoods, manipulates viewers emotionally and relies on questionable filmmaking techniques to create "facts" that support its point of view' (SeaWorld, 2017). In its conclusion to the '69 Reasons', the website declared that the claims 'that all killer whales in captivity are "emotionally destroyed" [. . .] are not the words of science' and that they are instead 'the words of animal rights activists whose agenda the film's many falsehoods were designed to advance. They reveal "Blackfish" not an objective documentary, but as propaganda' (ibid.). In this way, the resulting discursive battle for ownership of truth claims between SeaWorld and *Blackfish* about orca captivity brought both anthropomorphism and the norms of documentary into the debate. Using highly emotive language such as 'propaganda' and 'animal rights activists' and addressing critics and viewers with the probability that, if they succumbed to the film's message they had been misled and emotionally manipulated, SeaWorld challenged the legitimacy of the film's classification as documentary. The same strategy had been adopted three years earlier by the gas and oil industries in their attempts to debunk the film *GasLand* (2010), a documentary with an anti-fracking message credited with the consolidation of grassroots resistance to fracking on three continents. Attempts to undermine the legitimacy of *Blackfish*'s claims about Tilikum's emotional state by policing the limits of scientific discourse as well as building an implied opposition between 'science' and 'animal rights' thus aimed to redraw the lines around what should be regarded as truth claims. The pro-captivity SeaWorld discourse tried to debunk what it termed the 'misleading techniques of Blackfish' by recourse to the factual accuracies of nonanthropomorphic science. However, this strategy was difficult for the corporation to maintain when SeaWorld had relied on creating a highly anthropomorphised environment in which individuated orcas were ascribed with personalities and complex emotional and cognitive traits that paralleled those of humans, the cutification of cetaceans, exploitation of fantasies of the human-cetacean bond, and emotional appeals to audiences as key strategies to underpin commercial success. As one press article observed about the Blackfish effect,

> There is, of course, an irony to all this. As good as the film is, it could never have had such a profound effect on viewers if SeaWorld had not done such a good job of making the world fall in love with orcas in the first place.
>
> (O'Hara, 2017)

In this regard, SeaWorld had engineered the anthropomorphic conditions of its own demise and central to this was a narrative of mental suffering.

Damage to reputational capital has to be carefully managed by organisations. Where nonhuman animals are concerned public attitudes can quickly change (see Molloy, 2011). Alex Lockwood points out in his discussion of affect, 'public feelings can contribute to and maintain normative values within a culture; but they are also emotions that can be circulated by those same or alternative systems to challenge that culture' (Lockwood, 2016: 740). A positive corporate reputation confers significant advantages and in the case of entertainment companies this includes enhanced marketing opportunities, stronger customer bonding, loyalty, trust, and increased word-of-mouth recommendation. And the obverse of this is true, where poor strategic management of reputational capital adversely affects consumer relations. For a company such as Disney, the accrual of reputational capital is absolutely central to their strategic management and is, in large part, handled through their publicly perceived and actual corporate social responsibility (CSR) (Molloy, 2013). The company's 'green' credentials, links to conservation charities and environmental focus is absolutely central to the Disney brand and reputation. For a brand that is so closely aligned with cute anthropomorphised animals, most often shorthanded within popular culture to Bambi, what is crucial to the reputational capital mix is the public perception of Disney's relationship with and treatment of animals. In 2012, Disney instituted a 'use of live animals in entertainment policy' (Disney, 2012). This set out core principles around the safety of nonhuman animals and humans: that the use is respectful of the nonhuman animal; that educational and meaningful welfare or conservation components are integrated into the presentation of nonhuman animals; and that American Humane Association (AHA) guidelines must be followed with the addendum that the AHA tag line is desired in production credits (Disney, 2012). With occasional unspecified exceptions, it states that exotic live animals are not to be used outside of zoo/sanctuary habitat or natural environment. There is a no exception rule in the use of apes and other large primates. One outcome of this policy was that all the species other than the human protagonist in the 2016 adaptation of *The Jungle Book* are CGI.

The economic benefits of animal films continue to be balanced against the risks to reputational capital which impact on box office. Disney's *The Jungle*

Book (2016) used CGI technology to create all of the non-human species depicted on screen; the 2012 film *Life of Pi* used a mix of live and CGI animal performance; and *Rise of the Planet of the Apes* was heralded as a breakthrough for using only CGI at all stages of production. In these films, species that are being lost in the current mass extinction crisis – tiger, panther, gorilla, chimpanzee, orangutan (the big cats and great apes) – are being granted representational immortality via CGI. *Jungle Book* opened in 4,500 theatres in its first weekend. In other words, there were 4,500 representations of the Bengal tiger, Shere Khan, in circulation. According to World Wildlife Fund, there are only an estimated 2,500 real Bengal tigers left in the wild. As we do little to halt the decline of a species in the slippage towards extinction, we must ensure that we do not end up with only the fictions of liveness: a disturbing form of digital taxidermy. Moreover, other blockbusters such as *A Dog's Purpose* (2017), *The Revenant* (2015) and *The Hateful Eight* (2015) serve as a reminder that nonhuman animal labour, in these specific cases canine and equine labour, remains economically important to filmed entertainment where the risks to reputation are much reduced and there is less motivation for studios to replace horses and dogs with CGI representations. Nonetheless, the concern over reputational capital has forced some changes in the use of exploitation of nonhuman animal labour, and it is of note that the more focused the brand's association is with anthropomorphised nonhuman animals, the greater the risk of reputational damage in relation to public perception of an organisation's treatment of real nonhuman animals.

What can anthropomorphism do?

The discursive contest over truth claims about nonhuman animals all too often invokes anthropomorphism to bolster one side or the other. From a posthuman perspective, 'the capabilities of anthropomorphic activity to complicate simple binaries of human/nonhuman or human/inhuman, mean that anthropomorphism often emerges in discursive struggles to preserve ontological security and the taxonomic privilege of the category 'human'' (Molloy, 2001). Where anthropomorphism does confuse such binaries, animals are constructed as hybrid or liminal beings who are then subject to the push and pull of a speciesist politics (Molloy, 2001). As well as operating as a site that can be deployed to undermine the attribution of agency and subjectivity to nonhuman animals, anthropomorphism also has the potential to be an effective irritant to humanism and anthropocentrism, particularly where popular culture imagines what it is to be another animal, in ways that respect species difference but through a lens of similitude. Even where similitude is overexpressed this can result in creating a backdrop of care that has the potential to open up transformative opportunities. In addition, the narrative excesses that give rise to paratexts, a means by which commercial media enlarges the public conversation around a particular film, television programme and so forth, create the potential for resistant readings that can shift focalisation and reorient the discourse to consider the material realities of the life of an individual animal.

Popular culture mediates many of our encounters with other animals and in doing so it inevitably anthropomorphises. As Nik Taylor observes, 'anthropomorphism is unavoidable given that humans interpret the natural world and other animals (and indeed other humans) through their own embodied materiality' (Taylor, 2011: 265). The processes of mediation transform the modality of that interpretation such that it can move us to imagine from the nonhuman animal's point of view in ways that are not otherwise accessible. The mediated encounter is different to the material encounter between human and animal, but this does not automatically mean that the animal is wholly erased or their difference ignored. In this sense, narratives of nonhuman life and experience can engage us in empathetic connections even though mediated encounters equally bear the risk of normalising anthropocentric views or reinforcing other intersecting oppressive frameworks.

Anthropomorphism within popular culture is not without its problems. It is ironic that the contact points between science and the popular – natural history documentaries and wildlife films – are places where a gendered discourse of commodified anthropomorphism is still rife. Scientific discourse has attempted to eschew anthropomorphic language and disavow anthropomorphic practices but where science is mediated by the popular, anthropocentric frameworks of gender, morality and social norms are frequently used to engage audiences. Popular culture has proved useful in holding up a mirror to the ways in which our insistence on the privilege of human language can diminish the subjectivity of another sentient being, eroding their species-specificity. Taking anthropomorphism seriously gives us insight into the forms of labour that nonhuman animals perform, understood within a framework of capitalist exploitation. As such, the labour of performance and the exploitation of emotional labour are two of the ways in which nonhuman animals are economically and affectively productive for humans. These dynamics highlight the importance of thinking about anthropomorphism as a situated practice which can be and has been, exploited for capitalistic gain. Nonetheless, even when this happens, the normalisation of care for other animals can become an important mobilising factor. Anthropomorphism is enmeshed in human-nonhuman animal power relations and intervenes in discourses that shape the practices which govern the material lives of animals. In popular culture it engages both human empathy for and misunderstanding of nonhuman animals. It is therefore vital that we do not dismiss it out of hand but that we instead consider how it impacts on the material lives of all animals.

We all, humans and nonhumans, face challenging times. Climate change, animal agriculture as well as the various other forms of animal exploitation under capitalism result in the misery and suffering of billions each year. In the face of this it is vital that we find ways in which we can critique the current state of things with some level of pragmatism. Advocate and independent scholar, Kim Stallwood, writes about how, in the animal rights movement, pragmatism can be confused with a lack of commitment to advocacy. In the argument between incrementalism and abolition he calls for a view that balances 'utopian vision with pragmatic politics' and points out that while we might strive for an ideal, we must think about what is realistically achievable and there are times when compromise is the only realistic move forward (2014: 178). When we think about anthropomorphism the same must apply. To critique the speciesist and anthropocentric biases of popular media in favour of an ideal mediation of animal life and experience that does not anthropomorphise other animals is unrealistic.

Increasing urbanisation, biodiversity loss and changes to animal agriculture practices that have animals confined and contained mean that the encounters between human and animals are more likely to be through a screen. Sentimentalism and cuteness have come to dominate popular culture, the conditions of this mode and aesthetic differing from that of previous centuries. The neoliberal form of capitalism that we find today has produced new forms of precarity, social and economic imbalances and has marked out the shift from citizen to consumer and prosumer. Under these conditions, sentimentalism can be a salve, one that incorporates animals into the nostalgia for affective bonds of friendship and family and allows for identifications across species difference. Such affective appeals can function as a mobilising force and where anthropomorphism sits at the juncture between cute and sentimental there are opportunities for these sites to open up empathetic connections. At the same time, these sites are indeed the product of capitalistic enterprise that have utilised the labour of animals, but we must recognise that audiences are not uncritical consumers, prosumers have motivations other than economic gain, and media texts are polysemic. We arrive at these texts as embodied subjects who view but also feel. It is possible, I propose, that our engagements with mediated animals at these sites of anthropomorphism are not simply an exercise in casual anthropocentric narcissism; they may also be individual acts of being attentive to the lives and experiences of other animals.

Notes

1 This study looked at 59 public reviews on the Internet Movie Database (IMDB).
2 References to the real names were numerically equal, mentioned in almost one-third of the total number of reviews.

References

Batt, E. (2013) 'Op-Ed: SeaWorld's strange e-mail to film critics of Blackfish' *Digital Journal*, 15 July, online at http://www.digitaljournal.com/article/354375

Bradshaw, P. (2011) 'Project Nim' *The Guardian*, 10 August, online at www.theguardian.com/film/movie/142243/project-nim

Brooks, M. (2014) 'Five insights challenging science's unshakable "truths"' *The Guardian*, 29 June, online at www.theguardian.com/science/2014/jun/29/five-insights-challenging-sciences-unshakable-truths

Canby, V. (1993) 'Film boy and orca meet cute' *New York Times*, 16 July, online at www.nytimes.com/1993/07/16/movies/review-film-boy-and-orca-meet-cute.html?mcubz=0

Clave, S. (2007) *The Global Theme Park Industry*, CABI, Oxfordshire.

Cohen, M. (2014) 'The history of shark week: How the discovery channel both elevated and degraded sharks' *The Week*, 14 August, online at http://theweek.com/articles/444542/history-shark-week-how-discovery-channel-both-elevated-degraded-sharks

Dawkins, M. S. (2012) *Why Animals Matter: Animal Consciousness, Animal Welfare, and Human Well-Being*, Oxford University Press, Oxford.

Disney (2012) 'Disney's use of live animals in entertainment policy', 3 April, online at https://ditm-twdc-us.storage.googleapis.com/Disneys-Use-of-Live-Animals-in-Entertainment-Policy.pdf

Doyle, J. (2014) 'Sharks: The ultimate antiheroes of TV' *The Globe and Mail*, 9 August, online at https://beta.theglobeandmail.com/arts/television/sharks-the-ultimate-antiheroes-of-tv/article19974728/?ref=www.theglobeandmail.com&

Ebert, R. (1993) 'Free willy' *RogerEbert.com*, 16 July, online at www.rogerebert.com/reviews/free-willy-1993

Ebert, R. (1997) 'Free willy 3: The rescue' *RogerEbert.com*, 8 August, online at www.rogerebert.com/reviews/free-willy-3-the-rescue-1997

Ebert, R. (2011) 'Project Nim' *RogerEbert.com*, 6 July, online at www.rogerebert.com/reviews/project-nim-2011

Hogan, K. (2010) ' 'Free willy' star Bindi Irwin: I feel lost without animals' *People*, 23 March, online at http://people.com/pets/free-willy-star-bindi-irwin-i-feel-lost-without-animals/

Jones, C. (2017) 'SeaWorld attendance falls as it downplays killer whales' *USA Today*, 9 May, online at www.usatoday.com/story/money/2017/05/09/seaworld-sees-plunge-revenue-and-visitors/101459380/

Kaufman, A. (2013) ' "Blackfish" gives Pixar second thoughts on "Finding Dory" plot' *Los Angeles Times*, 9 August, online at www.latimes.com/entertainment/movies/moviesnow/la-et-mn-blackfish-seaworld-finding-dory-pixar-20130808-story.html

Klady, L. (1993) 'Review: Free willy' *Variety*, 6 July, online at http://variety.com/1993/film/reviews/free-willy-1200432813/

Klady, L. (1997) 'Review: "Free Willy 3: The Rescue" ' *Variety*, 8 August, online at http://variety.com/1997/film/reviews/free-willy-3-the-rescue-3-1117329825/

Klinger, B. (2006) *Beyond the Multiplex*, University of California Press, Oakland.

Lockwood, A. (2016) 'Graphs of grief and other green feelings: The uses of affect in the study of environmental communication' in *Environmental Communication*, Vol. 10 (6), pp. 734–748.

Luty, D. (2004) 'Free willy 3 the rescue' *Film Journal International*, 2 November, online at www.filmjournal.com/node/15151

Mendoza, D. (2013) ' "Blackfish" prompts school to cancel long-standing SeaWorld trip' CNN, 19 December, online at http://edition.cnn.com/2013/12/18/us/school-cancels-sea-world-california/index.html

Molloy, C. (2001) 'Marking Territories' in *Limen: Journal for Theory and Practice of Liminal Phenomena*, Vol. 1, online at http://limen.mi2.hr/limen1-2001/clair_molloy.html

Molloy, C. (2011) *Popular Media and Animals*, Palgrave Macmillan, Basingstoke.

Molloy, C. (2013) 'Nature writes the screenplays: Commercial wildlife films and ecological entertainment' in Rust, S., Monani, S., and Cubitt, S. (eds) *EcoCinema Theory and Practice*, Routledge, London and New York.

O'Connor, S., Campbell, R., Cortez, H., and Knowles, T. (2009) *Whale Watching Worldwide: Tourism Numbers, Expenditures and Expanding Economic Benefits, a Special Report from the International Fund for Animal Welfare*, prepared by Economists at Large, Yarmouth, MA, USA.

O'Hara, H. (2017) 'SeaWorld vs Blackfish: The film that introduced the world to the plight of Tilikum' *Telegraph*, 6 January, online at www.telegraph.co.uk/films/2016/06/04/seaworld-vs-blackfish-the-film-that-saved-the-whales/

Orlean, S. (2002) 'Where's willy?' *The New Yorker*, 23 September, online at www. newyorker.com/magazine/2002/09/23/wheres-willy

RottenTomatoes (n.d.) 'Critic reviews for *Free Willy*', online at www.rottentomatoes. com/m/free_willy

Rowley, T. and Molloy, M. (2017) 'Tilikum: SeaWorld killer whale from Blackfish documentary dies' *The Telegraph*, 6 January, online at www.telegraph.co.uk/news/2017/01/06/ tilikum-seaworld-orca-blackfish-documentary-dies/

Scott, A.O. (2011) 'Some humans and the chimp they loved and tormented' *New York Times*, 7 July, online at www.nytimes.com/2011/07/08/movies/project-nim-about-a-chimpanzee-subjected-to-research-review.html?mcubz=0

SeaWorld (2017) 'Why "Blackfish" is propaganda, not a documentary' *SeaWorld Cares* website online at https://seaworldcares.com/the-facts/truth-about-blackfish/#5

Shareholders Foundation (2017) 'Update in lawsuit for investors in shares of SeaWorld Entertainment Inc (NYSE: SEAS)' *Globe News Wire*, 6 March, online at https:// globenewswire.com/news-release/2017/03/06/931978/0/en/Update-in-Lawsuit-for-Investors-in-shares-of-SeaWorld-Entertainment-Inc-NYSE-SEAS-announced-by-Shareholders-Foundation.html

Snierson, D. (2017) 'Michael Phelps on racing a great white during shark week: "Sure why not?"' *Entertainment*, 11 July, online at http://ew.com/tv/2017/07/11/shark-week-michael-phelps-race-great-white/

Stallwood, K. (2014) *Growl*, Lantern Books, New York.

Taylor, N. (2011) 'Anthropomorphism and the animal subject' in Boddice, R. (ed) *Anthropocentrism: Humans, Animal, Environments*, Brill, London.

Tudor, A. (1989) *Monsters and Mad Scientists: A Cultural History of the Horror Movie*, Blackwell, Oxford.

WDC (2016) *A Guide to Responsible Whale Watching*, WDC, Wiltshire.

Wilmington, M. (1997) 'Without Keiko, third "Free Willy" does a belly flop' *Chicago Tribune*, 8 August, online at http://articles.chicagotribune.com/1997-08-08/entertainment/ 9708080051_1_keiko-whale-amusement-park

Yacowar, M. (2003) 'The bug in the Rug: Notes on the disaster genre' in Grant, B.K. (ed) *Film Genre Reader III*, University of Texas Press, Texas, pp. 277–295.

Index

Printed and bound by CPI Group (UK) Ltd, Croydon, CR0 4YY
01/11/2024
01782621-0017